Audel™

Managing Shutdowns, Turnarounds, and Outages

Audel™

Managing Shutdowns, Turnarounds, and Outages

Michael V. Brown

Wiley Publishing, Inc.

Vice President and Executive Group Publisher: Richard Swadley
Vice President and Executive Publisher: Bob Ipsen
Vice President and Publisher: Joseph B. Wikert
Executive Editorial Director: Mary Bednarek
Executive Editor: Carol Long
Editorial Manager: Kathryn A. Malm
Senior Production Editor: Fred Bernardi
Development Editor: Kevin Shafer
Production Editor: Pamela Hanley
Text Design & Composition: TechBooks

The author would like to acknowledge the contributions of Michael H. Bos, cofounder and former partner in New Standard Institute, in the development of the original version of this material. Additional assistance was provided via the generous time commitment and considerable talent of Alexandra L. Galli.

Contents

About the Author

Michael V. Brown is an Electrical and Maintenance Engineer with 30 years of experience in industry and has held positions at both the plant and corporate levels of Fortune 500 companies. As founding partner and President of New Standard Institute, he has designed and implemented maintenance management improvement programs for numerous industrial clients. New Standard Institute provides seminars, consultations, and computer-based training programs specific to maintenance-related subjects. For more than a decade, he has written articles that have been published in many U.S. and Canadian magazines, as well as on the Internet. Other books he has written and published by Wiley include *Audel Maintenance Planning and Scheduling* and *Audel Managing Maintenance Storerooms*.

Introduction

I first encountered the need for shutdown management when I worked in the chemical industry. My early experience was good. The first few companies I worked for placed a high regard on the downtime periods, so a lot of planning was put into the shutdown effort. When I started my consulting career, I soon learned that first-rate shutdown management was not the norm in many industries.

I had been providing planner/scheduler training and consulting for many years. A small part of our training included a discussion about shutdowns, turnarounds, and outages. It became evident that shutdown activities were treated with an alarming casualness at many of my client's plants. The cost for these shutdowns could amount to multiples of the annual budget for day-to-day maintenance. Equipment was disassembled before parts needs were identified. Facilities were shut down and work was started with absolutely no preplanning or estimates.

The beginning of the shutdown was accompanied by a lot of "pregnant activity": people waiting around for other people to start their work. This continued throughout the shutdown period. The result was a feeling of frustration. Very little of what needed to be accomplished was done, and the operation didn't run much better. In some cases, it ran worse.

As bad as it sounds, engineering and maintenance sometimes don't mix well. Maintenance workers often complain that they have to fix the problems that engineering has made. As a result, it's hard for them to believe that Engineering is incorporating methods that can help a maintenance effort. The use of project management tools is one such method.

Modern project management methods have provided a real benefit when it comes to scheduling shutdown events. Project management software (once rudimentary and awkward to use) has become more accessible and feature-filled. Electric utilities and petrochemical plants were the first adapters of this technology. Although the amount of software available was numerous, now most companies have zeroed in on a handful of programs.

I began to present a seminar called "Shutdowns, Turnarounds and Outages" around the U.S. and Canada about eight years ago. I also did some direct consulting in this area. My attendees and clients were persons from chemical plants, the petroleum industry, electrical generating plants, food manufacturers, and pharmaceutical companies. The seminar material (designed and improved based on attendee discussions and actual experience in industrial plants) has resulted in this book.

Chapter 1

Identifying the Work

Most maintenance departments have developed workable methods to handle day-to-day work. However, the most complicated call for maintenance resources comes when the plant is shut down for maintenance. Some plant operations shut down just because inventory is high or because business activity is low. Others are in a sold-out situation, but they just can't run the plant in its current condition. Government organizations sometimes mandate annual shutdowns for inspections.

Whether it's called a *shutdown, shut-in, downturn, turnaround,* or *outage,* the down period affords the maintenance department opportunities that may not be available again for a long time. The most demanding aspect of these periods is fitting a large compliment of work into a short period of time. It can be a period where the maintenance department really shines, or, as is often the case, these down periods prove the inadequacy of the maintenance department.

Some companies leave shutdown planning up to one individual or a separate group of people at the plant. Sometimes the planning of a shutdown is taken completely away from site personnel and handed over to a corporate engineering department. Plant personnel are queried for their suggestions on jobs to be performed, but laying out the schedule is left to the specialized group.

Gearing up for a major shutdown does not necessarily have to be relegated to a special group. A novice with some insight can coordinate a good shutdown. Using modern project management computer programs can help. These programs enable maintenance professionals to staff and coordinate the effort of hundreds of workers along with support equipment while minimizing the downtime and the costs involved.

Project management is a science that has developed out of necessity. The increasing cost of labor and materials has required tight controls on these resources to maximize productivity and limit losses. The increased complexity of manufacturing plants and the high cost of downtime has necessitated tight control over the execution phase to ensure that total costs of the project and accumulated downtime do not get out of hand.

Although project management has historically been applied to engineering and construction work, the tool of project management is uniquely adaptable to maintenance work associated with a shutdown.

Project management begins with the inception of the project and includes all phases of development. Following are the five traditional phases of a shutdown, turnaround, or outage:

1. Identifying the work to be accomplished and the scope of work that can be completed with the time and money allotted. The application of project management techniques hinges on how well this has been done.

2. Planning the work included in the project. During this phase decisions must be made as to how long each job will take to perform, how many people are required, what materials are required, and what special tools or equipment may be needed. Included in this planning are economic decisions of make vs. buy and in-house vs. contract decisions.

3. Scheduling of the project work. This phase determines the order in which work will be performed. Such scheduling efforts usually include some application of CPM (Critical Path Methods) or PERT (Program Evaluation Review Technique). This phase can employ computer programs that assist in the scheduling process.

4. Execution of the project itself. During this phase, communication among the participants on the status of work is paramount. Also, the means by which contingencies and additional or new work will be handled must be determined.

5. Reporting and documenting the shutdown activity. Shutdowns of operations are usually not common events. It's a good idea to document the shutdown preparation and execution. Any discoveries or needs for future work made during the shutdown should be documented.

This chapter begins the examination of shutdowns, turnarounds, and outages by taking a closer look at the first phase: identifying the work. The chapter then delves into risk management before concluding with a discussion on defining and limiting the scope of work to be done during a shutdown, turnaround, or outage.

Identifying the Work

Don't restrict the shutdown scope in the beginning. Try to identify all the work in the affected area. Efficiencies through good project management may provide an opportunity to do more work than initially envisioned. Development of a critical path schedule and subsequent load leveling often identify periods of low labor requirements. Extra work on the schedule can be performed during these periods. Even if these jobs do not require downtime, they can be performed when personnel become available. Only the available downtime and the money that can be prudently expended should limit the shutdown scope.

Identifying the work in advance is prudent, even if the required resources will not be available to do the work during the upcoming shutdown. Some future shutdown will provide an opportunity to perform this work.

This section examines the following critical areas when identifying work to be performed during a shutdown, turnaround, or outage:

- *Reviewing the maintenance backlog*—This includes reviewing preventive maintenance (PM) jobs, jobs that do not require a shutdown, history of equipment that is being affected, and records of predictive maintenance (PDM) work.

- *Identifying the preliminary work*—This involves checking such things as infrared scans, preparing checklists of tasks to be performed, identifying specific areas for inspection, checking with individuals who have experienced similar shutdowns, reviewing shutdown files, preparing plans for start-up activities, and preparing checklists for tasks to be done during the shutdown.

Reviewing the Maintenance Backlog

The current maintenance backlog will most likely contain work that requires a shutdown in the affected area. Some computerized maintenance management systems (CMMSs) allow these jobs to be pre-classified as shutdown, which makes them easier to find.

Utility equipment, outside the affected area, should also be considered during the search. Steam, compressed air, or electricity provided to the area can also be removed from service or curtailed to allow needed repairs to be performed.

Preventive Maintenance Jobs

Any search for preventive maintenance (PM) work to be performed during a shutdown should start with a search for overdue PM. Next, the planner should check all preventive maintenance work due to be performed before or after the shutdown date. It may be advisable to use the scheduled shutdown to change equipment oil rather than just add oil.

For example, assume the shutdown will take place on May 7 and the next major shutdown in this area will not occur for another three months. All PM work that will fall due during the three-month period after May 7 should also be considered for the May 7 shutdown. Also, some equipment lubrication can only be changed when the equipment is down. Be sure these jobs are on the shutdown schedule.

Jobs Not Requiring a Shutdown
Work in the backlog that may not require a shutdown should also be reviewed. As stated previously, resource load leveling may reveal idle periods for certain crews. Nonshutdown work can be performed for these crews.

Additionally, contract or specialized help not normally available may be present during the shutdown. An example of this could be an insulation contractor. If insulation work is required during the shutdown, the opportunity should be taken to do other insulating work while the contractor is on site.

Some nonshutdown jobs may require special preparation. Scaffolding, cranes, or other support equipment may be rented for the shutdown so this equipment won't have to be erected during a nonshutdown period.

Equipment History
A quick review of equipment history when planning a shutdown is always advisable. Basically, this review is centered on finding the equipment in the affected area that has cost the most to maintain, or has the most work orders charged against them. This historical search should be limited to work performed in the last year or since the last shutdown.

Many maintenance professionals feel that they don't need a sophisticated system to identify high-cost areas. This may be true in very small plants, but it is definitely not the case in larger plants. A detailed review of completed work orders is usually required to identify problem equipment.

In the days before CMMSs, copies of completed work orders were filed three different ways: by work order number, chronologically, and by equipment number. A maintenance professional hoping to find the equipment in the plant that costs the most money to maintain would simply look for the fattest equipment file. This method would help identify equipment that had the most work orders charged against them. The fact that numerous work orders are charged to one piece of equipment may not, in itself, be an indication of a problem, unless the majority of the jobs are for breakdown or emergency work. A more detailed review of work order costs may also be necessary to find problem equipment.

Today, the search for work to perform during a shutdown is much simpler. CMMSs store all work order data in one computer file. This data can be sorted and reviewed in more ways than is possible with a manual system. The equipment history can be tabulated and sorted, not only by number of work orders but also by cost.

A shutdown planner should pick out equipment that seems to have been particularly troublesome because of the high cost of repetitive or emergency work. Engineering changes or detailed inspections should be scheduled for this equipment during the shutdown.

Predictive Maintenance Records

Many companies today have a formal predictive maintenance (PDM) program. PDM programs for rotating equipment concentrate on vibration measurement and oil analysis.

Even companies with no formal PDM programs can benefit from a quick check of equipment condition during operation. An inexpensive, handheld vibration meter is all that is needed to make this check. Instruments that measure vibration velocity provide the best indication of the severity of the problem.

The Vibration Institute (a not-for-profit organization) has established levels of equipment health as a function of vibration velocity. In general, equipment that exhibits vibration of .3 inches per second velocity is said to exceed an acceptable vibration limit. This limit holds true on most industrial equipment operating between 600 rpm and 3600 rpm. Most formal program limits are based on this value.

If a formal program exists, the planner should check the vibration records for all equipment in the affected area that exceeds an overall vibration of .3 inches per second. If no formal program exists, the planner should acquire a vibration meter and take the readings. This data is truly worth the time spent.

If few downtime opportunities are made available on the equipment, it may be advisable to include equipment with even lower vibration readings (such as .25 inches per second). The assumption is that this equipment may be approaching the limit and the time is right to shut it down for repair or inspection.

Spectrum and phase analysis of the equipment with high vibration can determine the root cause of the problem. The most common sources found for high vibration are misalignment, imbalance, or bad bearings.

Oil analysis is another predictive tool that can provide valuable information about the serviceability of the lubricant, as well as the condition of the equipment in general. A spectrographic oil analysis can be performed by many oil suppliers or through an independent laboratory.

Spectrographic tests can detect wear metals such as aluminum, iron, nickel, chromium, lead, silver, copper, tin, zinc, or silicon. Elevated levels of these metals in the oil may not only indicate a problem, but can also be helpful in discovering the source. High nickel can be the result of rolling-element bearing wear. High copper can be the result of plain bearing failure.

The same test can identify the additives in the oil. If the additive elements and quantities don't compare to the specified additives that should be in the oil, contamination of the oil is suspected.

The physical and chemical tests performed can detect contamination or dilution with other oils. The tests performed include the following:

- Water percent (moisture in oil)
- Carbon buildup (indicating oil breakdown)
- Viscosity (oil too thin or too thick)
- Silica (dirt contamination)
- Total Acid Number (TAN) (indicates acids in oil)

Prepare to take oil samples on rotating equipment while the equipment is down (even if the oil is to be changed). Oil tests can provide results that indicate a deteriorating condition in the equipment. Changing the oil destroys that record.

Identifying Preliminary Work

As much as 20 percent of the total cost of shutdown can be accumulated before the shutdown work even begins. A number of prefabrication and inspection jobs may be identified for work that requires a shutdown. Purchases with long lead times should also be made early. Engineering and design work should be identified and started early.

The start of the project begins the day the preliminary work starts. The shutdown cannot begin if the prefabrication work is not complete or parts are not received. Some planners mark the start of the project with the first purchase order. The lead time quoted by the vendor or the time to complete the prefabrication or design work can all affect the start and subsequent finish time of a project. Anything requiring time prior to the actual shutdown is part of the total project.

Infrared Scans

One preliminary activity that can yield detailed information for the shutdown is infrared scanning. An infrared scan of electrical equipment is performed for the purpose of detecting loose or corroded connections on electrical equipment. Electrical current flowing through copper or aluminum conductors generates heat. This heat is mainly caused by the resistance of the electrical current flow that exists in the metal conductors. If a loose or corroded connection should occur, the resistance to current flow through the connection increases, and so does the generated heat.

An infrared scanner can detect this heat and can indicate the relative temperature. For example, a three-phase motor should have equal current flowing to all three phases of the motor, so corresponding phase connections should have the same temperature. If one connection is hotter than another, it may be loose or corroded. A problem is said to exist if the temperature difference is greater than 5°C.

Because of the expense of required equipment, an infrared scan is usually performed by an outside service. An electrician will most likely have to be scheduled to work with the contractor to open equipment panels and then close them after the scan is complete.

Following are a few items that should be considered before the scan:

- Prepare all electrical equipment for safe energized access during the scan. An infrared scan can only detect a problem on equipment that is exposed and energized. Some equipment access points and covers may not include hinges or handles. Install this hardware during a convenient shutdown.

- Identify all connections and equipment to be scanned with a detailed list. This will help to more effectively use the time of a contract scanning service.

- Schedule the infrared scan for a period in which load in the plant is near maximum. This will allow the worst problems to better exhibit themselves.

The electrical distribution system is not the only area where an infrared scanner can be used. Other problems that can be uncovered using an infrared scanner include the following:

- Kiln or furnace wall refractory failure
- Tube pluggage in a heat exchanger
- Insulation failure
- Leaks in roofing
- Other heating or cooling loss in buildings or machinery.

Identify Inspection Opportunities

Some work performed during the shutdown may not be immediately necessary but could provide some insight into future shutdown work. Exposed components provide opportunities for inspection and tests not normally available. High temperatures, chemical hazards, or radiation concerns may prevent inspection of equipment components and structures during operation. Once the equipment has cooled down or has been decontaminated, a more thorough inspection can be conducted. Consider the following examples:

- Collect dimensional information for future shutdowns
- Sketch up electrical wiring schemes that may be concealed while in operation
- Count exposed gear teeth in a gearbox (information that is very helpful in vibration programs)
- Perform nondestructive thickness tests on piping and vessels after the insulation is removed
- Perform eddy current tests on internal tubing
- Identify spare parts requirements on older equipment with no available manuals
- Take transformer or equipment oil reservoir samples that cannot be safely taken during operation
- Inspect and evaluate sleeve or bearing wear

Inspections such as the ones described here can uncover numerous problems. Shutdown personnel should guard against correcting each and every problem uncovered through a shutdown inspection. Unless the repairs can be accomplished very quickly during the shutdown or failure proves to be imminent, they should be left for future shutdowns. Making unplanned repairs as the problems are found diverts valuable time and labor that should be used for the planned jobs and could extend the shutdown period.

Solicit the Input of Others

When a problem arises during execution of a shutdown there is always someone around who predicted it or experienced the same problem before. These people just weren't heard early enough in the planning phase.

Success in shutdown planning lies in the details. Project or shutdown problems usually result from a lack of early information or responsibilities assigned to the wrong people. A shutdown planner would benefit greatly from the input of others in the plant.

Operations management can help identify additional needed work and potential problems by including engineering and maintenance supervisors. A major benefit can also be derived from soliciting the input from maintenance craftspeople, operators, line personnel, the storekeeper, purchasing, and other persons not normally included in the decision-making process. A structured group interview is one effective method of including these people.

The goal of a structured group interview is to document specific conditions and problems known to the attendees and to identify the solutions. These solutions are then prioritized.

Certain individuals may be stifled in a mixed group of management and hourly employees. Segregated groups should be considered, such as all operators or all maintenance mechanics. The group should contain a minimum of ten people and a maximum of twenty. The meeting should last no more than one to one-and-a-half hours.

Developing a question that best explains the problem to be solved starts the process. Questions such as "What can be done to make the next shutdown more successful?" usually elicit the most helpful responses. Avoid questions that may generate a laundry list of responses that will be difficult to prioritize. Questions such as "What jobs should be performed during the next shutdown?" should not be posed.

The rules for structured group interviews are very simple:

- *All responses must be presented in the form of a problem and a solution*—If a person has thought of a problem, he or she has most likely also thought of a solution to that problem.

- *All suggestions and solutions will be written down on a board or flip chart without rephrasing or debating them*—If people state only problems, explain to them, "It seems like you have thought a lot about this problem. Can you tell me of any solutions you have thought about?" It is alright to discuss the response after it is written down, but no one should be talked out of a response.

The steps to conducting a productive meeting are as follows:

1. Send out notices of the structured group interview and invite everyone. Provide a proposed schedule in the letter.

2. Everyone must be present before you start. When the meeting starts, no one else can be allowed to arrive in the middle of the session. Put a sign outside on the door stating, "Meeting already in progress—Please arrange to attend the next scheduled session."

3. Explain the purpose, goal, and rules to the group.

4. Write the question on a flip chart and post it in the front of the room.

5. Write down the responses (stated as a solution) to the question exactly as they are stated. Number them consecutively.

6. When a flip chart sheet is full, tear it off the pad and tape it on the wall.

7. When you reach 20 responses, stop listing responses.

8. Have everyone present vote for the ideas (posted on the wall around the room) by stating, "If you could vote for only five of these solutions, which five would it be?" Each attendee can go up to the posted sheets and vote by placing a mark next to the five responses.

9. Total the results for each item. Break ties between two or more items by asking for a show of hands, stating "If you had to choose 'A' or 'B', which one do you feel more strongly about?"

Individuals who have not been involved in brainstorming sessions in the past may be reluctant to be the first to talk. This is normal. After an initial response is given, group dynamics will take over and more responses will follow. Here are some tips that will help get that first response.

- Silence works wonders. After introducing the subject question, wait in silence. Fight your normal urge to break the silence. If you are talking, no one else can.

- If there is still no response, encourage attendees to consider a situation involving the subject question. "Think back over the last few weeks. Maybe you were in a situation that reminded you of our question. What was the situation and what kind of solution comes to mind?"

- Do not berate the group or put them on a guilt trip. Word will get around and any future sessions will be unproductive from the start.

- Do not coerce responses. Statements such as "We're not leaving this room until I have some responses" may get you a response or two, but the value of the session itself will be nil.

The result of a structured group interview will be a list of solutions to real and pressing problems in the affected area. These responses can only enhance the overall plan.

Reviewing Shutdown Files

Much of the work to be performed during an impending shutdown has been performed before. If the shutdown coordinator did a good job, the work performed in the past was well documented. Hopefully, a file of this work is available to the new coordinator.

The results of inspections or unfinished work are the most important information derived from this file. Any problems left or uncovered can be scheduled for rectification during the next shutdown. Any jobs that could not be started because of a lack of resources should also be considered for the new shutdown.

Additionally, mistakes of the past should not be repeated. Logistical errors and unexpected work stoppages can be eliminated in the next shutdown, as long as someone remembers these problems occurred. This information is often reported in a final shutdown report.

Identify Start-Up Activity

Make a list of activities that will occur at the end of the shutdown period. Cleanup, painting, touch-up, insulating, and start-up should be added to the list of shutdown work.

Sometimes unknown charges continue to accumulate against the shutdown after it is completed. Add a step to return rental equipment and compressed gas cylinders when the project is complete.

Shutdown Meetings

Meetings should be held throughout the planning phase of the shutdown. These meetings should be large at first (such as the structured group interviews described previously). Other brainstorming sessions may be helpful in the early stages.

Eventually, the project meetings should be restricted to a select group of people with a direct stake in the shutdown. Always publish an agenda before the meetings. Someone should be assigned to take the minutes of the meeting. These minutes should be published and distributed to all attendees and other stakeholders.

General Shutdown Checklist

Knowledgeable individuals should perform inspections of the affected area with the shutdown in mind. A checklist should be developed for this purpose if the shutdown is not a unique occurrence. All equipment and supporting structures should be inspected for the following:

- Missing guards
- Safety hazards

- Seal leaks
- Cleanliness
- Gage readings (compared to a limit)
- Oil levels
- Unusual sounds or smells (emanating from equipment)
- Deteriorating supports

Using a strobe light can enhance many inspections and checks. First, the condition of moving parts can be discovered. Additionally, the strobe light can be used to collect data from moving parts on rotating equipment. Finally, belt, coupling, sprocket, or sheave markings become visible under the strobe light.

The following items are common to many shutdowns and should be part of every shutdown manager's checklist:

- *Barricades*—Barricades should be considered to restrict movement of personnel for any of the following situations:
  - To limit entrance to, or egress from, any particular area of the plant or facility.
  - To restrict travel for contractors to and from their parking lot.
  - To protect all personnel from hazardous areas (or to minimize access to such areas) and to limit right-to-know training for all temporary personnel.
- *Building permits*—New construction or major improvements made during a shutdown may require permitting in some locales. Ensuring such legalities are covered in advance of the actual work should eliminate unnecessary and time-consuming delays.
- *Contractors' insurance certificates*—Most companies require minimum liability protection as well as proof of Worker's Compensation coverage for on-site contractors or other outside services. A file should be maintained for such certificates to minimize third-party litigation in the event of injury, death, or major damage.
- *Dust control*—The extra activity during a large shutdown can also be the source of excessive dust when unpaved areas are utilized as parking, staging, or even fabrication areas. Contracting a water truck service to regularly dampen down the areas can keep this problem in check, as well as improve relations with temporary personnel and the quality of work they

provide. Providing a temporary wash-down site for automobiles and trucks is also a recommended nicety.

* *Emergency showers and eye baths*—Extra emergency showers and eye baths should always be considered when the number of working personnel increases. These units are available on a rental basis with pressurized water supplies. The rental company can also be contracted to provide regular, documented inspection and testing. Request copies of such inspections for your own records.

* *Flag person or traffic control*—Services or individuals to control traffic or personnel flow should be investigated for large shutdowns. Consideration should be given to covering the following situations:

  * Exit to and from temporary parking areas onto local streets during shift changes.

  * Traffic control at heavily traveled or centrally located intersections within the plant or facility.

  * Special occasions for the movement of heavy machinery, cranes, arrivals of large shipments, or any extraordinary circumstance.

* *Liquid waste handling*—Liquid waste from certain cleaning operations may not qualify for handling within the in-plant industrial sewer. These materials need to be identified ahead of time for proper handling. If such handling is to be the responsibility of a vendor or contractor, review in detail the method of spill control, containment, and disposal.

* *Noise control*—Some shutdown operations may generate noise levels that are excessive. These operations must be identified ahead of time so that proper barricading or posting can be done.

* *Repairs of other damage*—Damage to existing structures or equipment should be assumed and accounted for when a large number of people are working in the same area. Additionally, all contracts with outside vendors and contractors should include repair clauses for damage to property fences, temporary facilities set up for such personnel, or other plant properties or facilities used by temporary workers.

* *Repairs of pavement*—Potential damage to pavement areas should be discussed with heavy equipment contractors ahead of time. If load-bearing capacity is unknown, plant roadways should be tested. Contractors should be advised of areas

where damage is probable and kept from movement in such areas.

- *Scaffolding*—Be sure to put up all scaffolding beforehand. If several contractors or scaffold rental agencies are to be used at the same time, require that each mark their own scaffolding so that it is not confused with others. Scaffolding is often moved from site to site during a shutdown, and the probability of mixing is fairly high. Requiring a different color marking from each supplier will help to keep it all identifiable.

- *Solid waste handling*—As with liquid wastes, potential handling problems can exist for solid wastes, especially when hazardous classifications are involved.

- *Supervisory coverage* (dark shifts and weekends)—There should always be a company representative any time temporary personnel (not employees) are in the plant. This individual is responsible for adherence to safety rules and represents the company in the event of an injury or incident.

- *Temporary buildings and enclosures*—Temporary buildings and enclosures are often the direct responsibility and cost of vendors and contractors. It is advisable to review with each supplier of temporary structures the following areas of coverage:

  - *Temporary cafeteria or eating facility*—Ensure that some provision is made, including vending equipment. Work through the logistics of restocking vending equipment (when it will be done, which supplier will be used, and so forth).

  - *Temporary first aid*—Large contractors should provide their own licensed Emergency Medical Technician (EMT) or First-Aid Technician along with a facility for primary care.

  - *Temporary heat and light*—Temporary parking areas used during 24-hour shutdowns should be provided with adequate lighting. Test the lighting before starting the shutdown.

  - *Temporary showers and change rooms*—Some shutdown work may necessitate the need for clean and dirty change rooms and shower facilities. The need for (and provision of) such services should be handled before the shutdown begins.

  - *Temporary storage*—Storage for material, tools, and equipment should be the responsibility of the vendor or contractor. Security for such storage and liability if theft or damage occurs should be determined before any material, tools, or equipment comes on site.

* *Temporary telephone*—Temporary telephones should be brought into the plant. These should be located in the normal temporary break areas. It is the responsibility of the vendor or contractor to ensure that abuse of this equipment does not occur.

* *Temporary toilets and water*—Portable toilets and potable water stations should be brought into the temporary structure area. If these facilities are to be staged within the plant or facility proper, it is advisable to arrange ahead of time how and when they will be serviced.

* *Temporary power*—If an unusually large contracted work force is expected, the utilities to such a camp town may tax existing capacity. Identifying the potential need and providing a temporary source from the local utility is advisable.

* *Smoking areas*—Set up smoking areas ahead of time or else contractors will be smoking on the job (which may be against plant rules).

* *Temporary construction protection*—Temporary protection of equipment undergoing maintenance or construction in progress should be addressed. Following are typical areas to be considered:

  * *Large machinery in overhaul*—Protection of bearings, machined surfaces, and tolerance items must be considered.

  * *Areas requiring special attention*—Protection of concrete forms, flange faces of prefabricated piping, and delicate instruments are just a sampling of items that should receive specific attention.

* *Gang locks and shift locks*—The Occupational Safety and Health Administration (OSHA) allows the use of gang locks and shift locks as long as adequate procedures and controls are in place to ensure that such locking devices really provide the necessary protection. It is strongly advised that lockout procedures be reviewed in advance, especially where large numbers of workers are involved or many different outside companies are on site at one time.

* *Make up a blind list*—Know where every blind is installed, and make sure that each is removed prior to start-up.

Work Lists versus Work Orders

The result of the many meetings and inspections will be work lists. Many companies start a shutdown with just these lists and then realize there is no document for charging labor and/or withdrawing materials from the storeroom.

Work lists should be converted to work orders so costs associated with the work can be accumulated. Any purchases that may not be directly associated to labor costs should also be covered on a work order as well as a purchase order. Relating all purchases to a work order streamlines the final accounting of the project. Define any engineering required by assigning a work order for this activity as well.

Any remaining activities that cannot be converted into work orders (such as operations decontamination or safety inspections) should be clearly listed for inclusion in the total project schedule.

Once a solid work list has been developed and finalized, we need to address *what can go wrong* with these jobs. Even the best-planned job can go awry during shutdown execution, and unforeseen situations can occur that must be addressed. These situations can be mitigated once the shutdown team goes through a risk management process.

Risk Management

All projects include some element of risk. It's natural to be optimistic that you will overcome any unplanned event during the execution of a project. It's also normal to be anxious that all things that can go wrong will go wrong. The adage "optimism blinds, pessimism paralyzes," applies very well to projects. Project risk management strives to achieve a balance between optimism and pessimism by confronting potential risks during the planning phase of a project.

The Project Management Institute (PMI) defines risk management as a subset of project management with the following four basic components:

- Risk identification
- Risk quantification
- Risk response development
- Risk response control

This section examines each of these four components in detail.

Risk Identification

A process should be set up to identify the risks that exist with a project. The development of a list of risk situations is a good team exercise. Often, the insights of one member will stimulate the thinking of others. The following items can be explored to stimulate discussion and flesh out hidden risk:

- *Staffing assumptions*—Some important activities may depend on the attendance of essential personnel. Identify people who will be indispensable during project execution because of their special skills or knowledge. Next, determine whether or not the project can proceed without these people. Also consider the potential for work stoppages or slowdowns.

- *Estimate risks*—Identify time and cost estimates that were developed with minimal information.

- *Procurement problems*—Any deliveries expected during the project execution should be reviewed for potential delays or even cancellation. Items from sole source suppliers represent the highest risk.

- *Project files*—Review previous project results. Even shutdowns performed in a different area can provide some insight into potential problems.

- *Commercial data*—Review trade articles for insight into some problems that others have encountered. The American Society of Professional Estimators (ASPE) has published articles on shutdowns and major modifications performed in industry.

- *Project team knowledge*—Query the project team. Team member recollections of previous projects are useful, yet less reliable than documented results.

- *Possible weather conditions*—Major storms (or just rain and snow) can affect the schedule of a project or shutdown.

- *Nature of the project*—Sometimes the magnitude of internal damage is an unknown. Corrosion, abrasion, or wear may be higher than expected. A similar risk exists when primary activities of a project will be performed for the first time. Application of new technologies or methods falls into this category.

Risk Quantification
The list developed from the risk identification process will most likely exceed the capability to mitigate all potential problems. The items on the list must be quantified to temper any response plan. The following two questions should be asked of the project team for each item on the list:

- Is the probability that this risk will be encountered high or low?
- Would an occurrence of the event significantly lengthen the project?

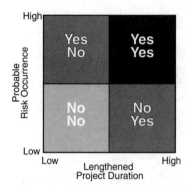

Figure 1-1 Quantifying the Risk.

Figure 1-1 can be used to weight your reaction to these two questions.

- *The Yes-Yes Quadrant*—The risk situations that fall into the Yes-Yes quadrant are too high to be ignored. Necessary steps and associated costs to minimize the risk should be built into the project plan.

- *The No-No Quadrant*—The risk situations that fall into the No-No quadrant should be ignored. The potential of occurrence is minimal and, even if the situation develops, the impact on the project schedule is negligible.

- *The Yes-No and No-Yes Quadrants*—The remaining two quadrants are economic issues. An assessment must be made of the cost to mitigate the risk situation to quantify these situations. If the cost is minimal, including the solution into the project plans would be good insurance.

Risk Response Development

Responses to a potential risk event fall into the following three categories:

- *Avoidance*—Eliminate the possibility of an occurrence
- *Mitigation*—Reduce the monetary expense of the occurrence
- *Acceptance*—Live with the consequences of an occurrence

Avoidance of problem usually costs money and can extend the project duration. For example, the use of a backhoe to dig a trench may damage buried piping. Consider paying the extra money and taking the extra time to hand-dig the trench.

Mitigation is the same as taking out insurance against the problem. For example, the simple act of opening or testing a component can damage it. An A-C high potential test of electrical cable can destroy the cable if it is in poor (but still operable) condition. Plans should be made to replace any cable that fails because of this test.

Acceptance of a problem means living with the extra down time and higher project execution costs. This is reasonable if the cost to mitigate the problem is about the same as the cost of the occurrence.

Risk Response Control

Executing the risk management plan will obviously begin when an identified risk event occurs. However, even the most thorough review of potential events cannot identify all risks, so a plan of action may have to be implemented on the fly. This plan must include the cycle of identification, quantification, and response. The key is in data collection, as shown here:

- Determine the information contained in the data that team members are reporting.
- Don't play hunches. Get the facts, and then act accordingly.
- Delegate (but monitor) action steps to reduce the effects of the event.

Defining and Limiting the Scope

Definition of the scope should begin with a statement about the purpose of the project. It is best if this statement is put in terms of a problem to be solved. A couple of good examples are "This shutdown will resolve questions about the condition of the #5 Boiler" or "This shutdown should build back capacity into the #2 Packaging Line."

Identify the stakeholder in the project. Who wants the shutdown and who will benefit from it?

- The maintenance department may want the shutdown to perform needed repairs.
- The operations department may want the shutdown because inventory levels are low or product quality has dropped noticeably.
- The engineering department may want the shutdown to make a modification or new installation.
- The plant may have to be shut down because of operating license requirements (such as boiler inspections).

Defining the Constraints of the Shutdown

Most projects have three basic constraints: elapsed time, resources, and money. A goal for the project coordinator is to determine which constraints are most important. *Elapsed time* is usually the most important constraint for industrial shutdowns. A specific amount of production downtime may be allotted for the work to be performed. If the work cannot be performed within the designated period of time it may have to be scaled back.

A *resource* refers to materials, tools, and equipment, but mostly labor. There is an upper limit to the labor hours available at any time. Even though the work force can be augmented with contracted help, this resource may not be adequate for all work to be performed. Particular jobs may require specialized skills that are in limited supply. Also, there is a limit to the number of people who can be working in close proximity to each other.

Money is usually the last constraint considered. The assumption is that most of the work must be done eventually, so why not now. The cost is inconsequential. However, if this attitude persists, the costs could run away because of overtime and penalty charges, putting the continued operation of the plant in jeopardy.

An old saying states, "Time is money." There is a cost of downtime that must be considered. For example, electric utilities must balance the cost of downtime against project costs. Staffing levels and the scope of work to be performed must be considered together with the lost revenue from not generating. Some utilities must pay a penalty for not being connected to the grid.

Quality may be a fourth constraint that is often overlooked. The quality of the work performed is directly related to the long-term success of the shutdown. If workmanship is poor, the shutdown will have to be repeated sooner. Usually, the quality of the work performed depends on the time provided to perform the work. However, experience and skill play a big part as well. Assigning the right people to the right job at the right time is one key to getting the best job possible.

Prioritizing the Proposed Work

After the initial scope of work has been defined, some prioritization is necessary. Most project management software programs allow you to establish a relative priority (along with the precedent activities) for each job. For example, one program initially sets the priority of every job at 500, with a possible range of 0 to 1000. The higher the number, the higher the priority. This prioritization should take into account the following factors:

- The future impact on capacity if the job is not performed
- The probability the problem will worsen before the next shutdown
- The increased cost of the job if it is delayed until a future date

All potential jobs should be entered into the project management program—even the ones with the lowest priority. Once the priorities are set and precedent activities are determined, the three project constraints (elapsed time, resources, and money) should be evaluated. A lower priority job should be dropped if any of the constraints are exceeded.

Preliminary Estimates of Proposed Work

After the initial activity of defining the scope of work is completed, you must develop preliminary cost estimates of the proposed work. These estimates are necessarily rough, attempting only to assign raw labor and material costs. Often, historical records of similar work performed on prior shutdowns will provide quick estimates at this stage. On other jobs (for which no historical records relate), an armchair analysis is the best means for establishing preliminary cost estimates.

Establishing a meaningful estimate for material costs may be difficult for some jobs. A typical ratio of labor to material costs is often used in this case. The assumption is made that any job would be typical of other maintenance work and would require an average amount of material. For example, assume a particular job requires an estimated 120 hours of maintenance labor to complete at the going internal rate of $35.00/hour. The Labor costs would be $4200 (120 hours × $35.00). If the ratio of labor to material costs has historically been 1:1.2, then the material costs can be estimated to be $5040 ($4200 × 1.2). The total job cost would be $9240 ($4200 + $5040).

Maintenance Shift Efficiencies and Cost Factors

In developing shift schedules, some planners erroneously assume that shifts of differing durations will return comparable hours of real work time to the project. This is not necessarily true. In many plants or facilities, some additional loss occurs when 10-hour shifts are scheduled. Even though 2 hours have been added to the workday, only $1\frac{1}{2}$ hours are available for additional work. For twelve-hour shifts, the loss can be even greater. In many labor contracts, the 12-hour shift allows the maintenance worker to request a meal break after 10 consecutive hours have been worked. This meal break, though contractually limited to a half hour, can easily increase to an hour in length, as the workers at this point are exhausted. In reality, the two additional hours added to the workday from the 10-hour schedule only add one half hour of real work time to the day. This breakdown can be seen in Table 1-1, which shows a single workday as it compares to various schedules.

Table 1-1   Single Workday Compared to Various Schedules

Shift Duration (Hours)	Days per Week	Schedule Hours	Available Hours	Available/Schedule Ratio
8	5	40	40	1.0
8	6	48	48	1.0
10	5	50	47.5	.95
10	6	60	56	.93
12	5	60	50	.83
12	6	72	60	.83

The effectiveness of the maintenance work force on continuous schedules (7-day weeks) should be considered when developing work schedules. The physical demands of a continuous work schedule become very apparent after the first week of the shutdown. Fatigue takes its toll as the second week progresses. Many plants apply the factors shown in Table 1-2 to the available work hours as the shutdown moves into subsequent weeks.

Table 1-2   Availability Factors

Shift Duration	Second Week	Third Week	Fourth Week
8	.9	.8	.73
10	.85	.73	.61
12	.80	.64	.51

In applying the factors, the labor hours that are projected to be scheduled are derated by the factor, and those hours are the hours that are, in fact, scheduled for execution. For example, if there were 1000 hours per week available on an 8-hour shift schedule, only 900 hours would be scheduled for the second week, 800 hours scheduled for the third week, and 730 hours scheduled for the fourth week.

Cost factors should also be taken into account when developing shutdown schedules. Work hours after 8 hours in any given day, or work hours in excess of 40 hours in a given week, are paid at a premium of $1\frac{1}{2}$ times the base hourly rate. Many plants also pay a double-time premium for work performed on a seventh consecutive day. Table 1-3 shows the relative hourly rate for varying work schedules over one week.

On the basis of the previous data, most plants utilize 10-hour shifts scheduled over 6 day workweeks. For work that must proceed

**Table 1-3 Relative Hourly Rate for Varying Work Schedules
over One Week**

Shift Duration (Hours)	Days per Week	Available Hours	Straight Time Hours	Hourly Cost Ratio
8	5	40	40	1.00
8	6	48	52	1.08
10	5	47.5	55	1.16
10	6	56	70	1.25
12	5	50	70	1.40
12	6	60	88	1.46
8	7	56	68	1.21
10	7	59.5	90	1.51
12	7	70	112	1.60

continuously, a staggered shift schedule is arranged, with a portion of the work force working a Monday through Saturday schedule, and the rest of the work force scheduled on a Tuesday through Sunday schedule. The importance of communication from shift to shift is critical in such a schedule.

Comparing Preliminary Estimates to Budget
After preliminary estimates have been made and, if necessary, massaged to reflect shift efficiencies and costs, the costs can be totaled and compared to the budget that has been approved for the shutdown. Shutdown costs are usually a major expense in any plants' operating budget. An early indication of what these costs might be is vital.

Hopefully, the estimated total cost of anticipated work does not exceed the budgeted amount. If this is true, more formal planning can now be done to more clearly define the resource requirements and cost of the jobs.

The total of the preliminary estimates may exceed the approved budget amount. In such instances, the early prioritization of the proposed work can be used to consider candidates to be dropped from the project scope. Since preliminary estimates will now be available on major jobs, an economic analysis can now be made in which the preliminary cost can be weighed against the potential benefit to the plant.

If the total of preliminary estimates cannot be reduced to within budget allowances, upper management must be made aware of the dilemma and brought into the decision-making process. Maintenance planners should not attempt to make these decisions alone, nor should they blithely proceed with project development hoping that preliminary estimates are overstated. The idea that all the proposed work can be accomplished at or under budget allowances must also be looked out for.

Determining Contract Work

In considering the amount of work to be performed in a typical shutdown, it is usually a given fact that the internal maintenance resources will be greatly overtaxed if they are the only considered work force. It is inevitable that some (if not a major portion) of the planned work will have to be performed by outside contractors or vendors. Contract or vendor supplied work is often the best alternative for the following situations:

- *Technical support*—Work that is very technical or specific to a manufacturer or special machine is often best contracted out. The unique skills, tools, or methods are often outside the technical scope of the maintenance personnel, or the skills are used so infrequently that retention is difficult. Typical of such work would be turbine/generator inspections and repairs during a unit overhaul for an electric utility. Industrial masonry work (such as refractory linings) is often contracted to experienced brick-laying companies.

- *Nontechnical support*—Work that is mundane (or requires relatively low skill levels) is often contracted out. In such instances, the skill levels of the maintenance personnel are much better applied to shutdown work requiring their knowledge. Typical of such jobs would be labor-intensive tasks of cleaning operations, catalyst removal and addition, or removal of abandoned equipment and systems.

- *Work that can be performed off-site*—Work that can be performed off-site is best contracted out. Such work will not place additional burden on plant parking, facilities, or other ancillary plant services. Fabrication, overhauls, testing and calibrations are examples of off-site work.

- *Work requiring special equipment*—Work requiring special equipment is almost always contracted. Since the special equipment is needed only during the shutdown period, it does not make sense for a plant or facility to own such equipment for the infrequent usage. Pressure blasting, nondestructive testing, and aerial painting are typical of contracted work.

Determining Internal Labor Demands

After considering work that can be contracted out, the remaining work in the initial scope must be performed by the internal labor resource. Using the initial preliminary estimates and proposed work schedules, an estimate can be made to determine if the existing maintenance work force can perform the work. If the existing work force is fully scheduled, some additional scope must be considered for performance by an outside contracted resource. It is well advised to schedule the existing work force to no more than 90 percent of the available maintenance hours. Some contingency must remain to allow for emergency situations elsewhere in the plant and possible reductions in the work force because of sickness, absenteeism, or other factors.

Job Input Cut-Off Date

Job input must have a cut-off date. This is usually a minimum of two weeks before the start of a small shutdown but could be as much as one or two months for a very large shutdown. The remainder of time before the projected start of the shutdown is needed to develop preliminary estimates, compare projected work to the approved budget, decide which work will not be done if the preliminary estimates exceed budget allowances, determine which work will be contracted, and finalize the project schedule.

It is imperative that management support the cut-off date. Even upper management should not have the ability to submit additional work after the cut-off date unless there are mitigating circumstances.

Prioritizing Last Minute Requests

It is also a good idea to decide how new work will be handled. New work consists of repairs identified after the start of the shutdown. Some thought must be put into qualifying this work. What criteria will be used to justify new work displacing work originally included in the shutdown? What will justify the displacing of work from the original scope of work? These issues are usually handled by the economics of the shutdown.

Maintenance should guard against giving undue preference to new work identified in the course of the actual shutdown. It is the tendency of many managers to overreact to this work and give it more that its due. In return, critical work first considered in the original scope is displaced from the schedule and subsequently causes reduction in capacity or additional downtime before the next scheduled shutdown.

Summary

The most complicated call for maintenance resources comes when the plant is shut down for maintenance. The most demanding aspects of these periods are fitting a large complement of work into a short period of time. It can be a period where the maintenance department really shines, or, as is often the case, these down periods prove the inadequacy of the maintenance department.

Project management is a science that has developed out of the realities of increasing cost of labor and materials. Although project management has historically been applied to engineering and construction work, the tool of project management is uniquely adaptable to maintenance work associated with a shutdown. Shutdown steps include identifying the work, planning the work, scheduling the work, execution, and reporting/documenting activity.

We should not restrict the shutdown scope while identifying the work. Development of a critical-path schedule and subsequent load leveling often identify periods of low labor requirements. Extra work on the schedule can be performed during these periods.

Identifying the work involves reviewing the maintenance backlog, preventive maintenance jobs, as well as jobs not requiring a shutdown to take advantage of resources and equipment available during the shutdown. Additionally, equipment history stored in a computerized maintenance management system (CMMS) should be checked for hidden problems that can be remedied with scheduled shutdown work. Predictive maintenance records (such as oil analysis and vibration monitoring results) will pinpoint equipment that would benefit from shutdown work.

It is also necessary to prepare for the shutdown period by performing some preliminary work. As much as 20 percent of the total cost of shutdown can be accumulated before the shutdown work even begins. Prefabrication and inspection jobs may be identified for work that requires a shutdown. Purchases with long lead times should also be made early. Engineering and design work should be identified and started early.

An infrared scan of electrical equipment is one preliminary activity that can yield detailed information. Prepare to perform the scan by ensuring safe access to the equipment while it is energized. Identify all the connections and equipment to be scanned with a detailed list. Schedule the scan for a period in which load in the plant is near maximum.

Other problems that can be uncovered using an infrared scanner include kiln refractory, furnace refractory, tube on equipment (such as heat exchangers), thermal insulation, building heating loss (roofs, walls, and so on), and machinery components (such as bearings).

Inspection opportunities during the shutdown should also be identified. Collecting dimensional data, electrical circuit sketches, gearbox inspections, nondestructive tests on empty insulated tanks, eddy current tests, hard to get oil samples, as well as evaluating sleeve or bearing wear all are important.

Solicit input for the upcoming shutdown from maintenance workers and operations personnel. A structured group interview is the best forum to get this input. Try to mine solutions from the interviews rather than a list of potential problems.

Mistakes of the past should not be repeated. Develop a regimen of established shutdown files. These files can be reviewed prior to the next shutdown to reveal the results of inspections or unfinished work. Make a list of activities that will occur at the end of the shutdown period. Cleanup, painting, touch-up, insulating, and start-up should be added to the list of shutdown work.

Meetings should be held throughout the planning phase of the shutdown. Shutdown meetings should start out large and include many people but should eventually be restricted to a select group of people with a direct stake in the shutdown.

Build a shutdown checklist, specific to your site. Facility walk-throughs should be organized to find obvious problems (such as missing guards, hazards, and leaks). A detailed list of items that may create a problem during shutdown execution should also be developed. The need for barricades, building permits, insurance certificates, dust control, and so on should be listed for the current shutdown and future use.

All work lists should be converted to work orders. Costs associated with the work can be accumulated in a CMMS only when this occurs. Work such as operations decontamination or safety inspections should also be converted to a work order, even though hours may not be charged against this work.

Risk management as defined by the Project Management Institute includes risk identification, risk quantification, risk response development, and risk response control. The project manager will have clear plans on how (or whether) to mitigate potential problem that may come up during a shutdown once a risk management process has been completed.

Defining the constraints of the shutdown involves reviewing elapsed time, resources, and money. A fourth constraint of quality (that is, workmanship) can also be considered. These constraints should be used to start prioritizing the proposed work. Preliminary estimates of proposed work should also be developed to determine the scope of the shutdown.

How and when to use a contractor is always a question prior to a shutdown. Contractors can be used for technical support, general labor, off-site work, or work requiring special equipment.

A cut-off date should be established for all work that may be added to the shutdown schedule. This is usually a minimum of two weeks before the start of a small shutdown but could be as much as one or two months for a very large shutdown.

Chapter 2 discusses the methods to plan shutdown work, the idea of work packages, and how to deal with contracts for services.

Chapter 2

Planning the Work

Having identified the work to be accomplished and the scope of work that can be completed with the time and money allotted, the next phase in management of a shutdown, turnaround, or outage is the planning phase. This phase entails deciding how long each job will take to perform, how many people are required to perform the work, what materials and equipment may be required, and whether in-house or contract resources will be used.

This chapter examines the planning phase of the project. Here we discuss the following major topics:

- Planning the shutdown
- The planning thought process
- Estimating tips
- Work packages
- Contracts for services

Planning the Shutdown

It takes time to prepare for a successful shutdown. One of the biggest mistakes people make when planning a shutdown is not allowing enough time to plan it. The time required to prepare for a shutdown depends on the number of labor hours required for the work involved. A quick estimate of all contract and site labor requirements should be added together and compared.

Figure 2-1 shows the range of time required to prepare for a shutdown. Assume that a 100-person labor force will take part in the shutdown. Also assume the shutdown will only be staffed 8 hours per day for 5 days. The total labor of 4,000 hours will be required for the shutdown. From Figure 2-1, you can see that between 100 and 450 planner hours will be required to prepare for the shutdown. The lower figure corresponds to a shutdown that has been performed in the past and is largely routine. The higher figure corresponds to the time required to plan the shutdown from scratch.

The lead-time required for the example shutdown is between $2\frac{1}{2}$ weeks to $2\frac{1}{2}$ months, assuming one planner will be working exclusively on shutdown preparation. Of course, this time can be cut in half if two planners are assigned to the project. The time to prepare could also be cut by involving other people in certain parts of the planning phase.

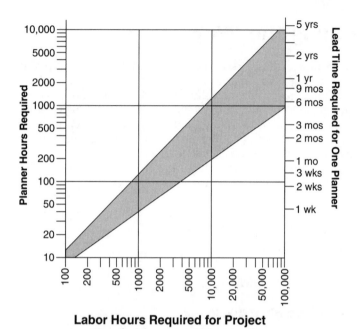

Labor Hours Required for Project

Figure 2-1 Project Labor versus Planning Time.

The Shutdown Organization

The first step to setting up a shutdown organization is choosing the coordinator. A successful shutdown coordinator should fully comprehend the scope of the project. Someone else should be chosen if the major work to be performed is outside the expertise of an individual being considered. The shutdown coordinator should also be someone who works well with people and can make decisions.

Next, the organization should be developed. Any existing organizational structure devised for day-to-day activity in a company is wholly inadequate for the coordinated activity required during a shutdown. For example, information not normally required by the maintenance department (such as product inventory levels or raw material deliveries) becomes exceedingly important to the timing constraints of a shutdown.

Additionally, the departmental hierarchy does not respond quickly enough to problems encountered during shutdown execution. Day-to-day operation of a facility is often likened to a sports team's effort. Sports teams practice and fine-tune their skills. The activity during a shutdown usually does not allow the luxury of practice. Shutdown requirements are more akin to the need-to-know requirements of a military organization. A limited number of people should be allowed to make major decisions, and those decisions can only be made with the blessing of the shutdown coordinator.

A traditional, top-down hierarchy is usually the best organization for a shutdown, as shown in Figure 2-2.

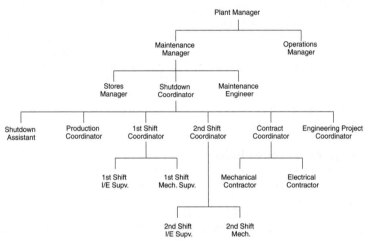

Figure 2-2 Shutdown Organization.

The traditional structure is shown at the top, with the operations manager and the maintenance manager reporting to the plant manager. Below the maintenance manager are the stores manager, maintenance engineer, and shutdown coordinator. There can be what is called a *dotted-line connection* between these three individuals.

Below the shutdown coordinator is the shutdown organizational structure. The shutdown coordinator may require an assistant to help develop schedules and reports and to perform some of the accounting functions. A production coordinator position should be assigned by the operations department to act as a liaison for production-related information. This individual can also coordinate the safety inspection and lock-out of operations equipment. Two shift coordinators are shown, assuming the shutdown will be scheduled over a number of shifts. All first-line maintenance supervision will report to the shift coordinators.

A contract coordinator may be required if contractors are going to be employed during the shutdown. This individual will communicate all instructions from the shutdown coordinator to contractor supervision.

Finally, an engineering project coordinator should report to the shutdown coordinator. Any projects or modifications that go on during the shutdown may affect the completion of the shutdown. As a result, the engineering project coordinator should communicate all progress of these jobs and should be informed if the completion of these jobs will be interrupted by other work.

The shutdown organization is geared toward the execution phase of the shutdown. Another structure may exist for the planning phase of the shutdown in which certain individuals are assigned jobs to plan and expedite. A formal structure may not have to be developed for the planning phase, but the Shutdown Coordinator should be in constant contact with all marginally involved individuals. Additionally, these individuals should be required to report on the status of the jobs they had originally been assigned to plan, even if they are not directly supervising the jobs.

Assigning Responsibility

As the scope and limitations of the shutdown become known, it may become clear that the maintenance planner does not have the time to properly plan and follow up on every activity. It is at this point that assignments should be made to other individuals in order to spread out the responsibilities. Table 2-1 shows an example of such an assignment list.

The list of work should contain all the jobs required to be completed during the shutdown. The assigned individuals should be informed (in writing) that they are responsible for the job plan and estimate, as well as for a material list and special tools or equipment list. These individuals are also responsible for purchasing any material, renting any special equipment, and hiring contract labor for the job.

Individuals assigned this responsibility should also follow the execution of the job and should be the liaison between the field activity and the shutdown planner. These individuals may not have direct supervisory responsibility for the jobs in their charge but should at least know the status at any moment during the project.

This is not to imply the planner should simply delegate all jobs and responsibilities. If progress reports are not forthcoming, the planner should seek out the individuals in charge. The planner should check the assumptions and estimates of these individuals to ensure that they are on the right track.

Parts, Material, and Equipment

Labor and engineering can all be acquired if they are missing during the shutdown (usually at elevated cost). Parts, material, and

Table 2-1   Scope of Work—C-101 Shutdown

Description of Proposed Work	Responsibility
Decontaminate, open and inspect C-101, repair damaged trays (if any)	Operations
Replace C-101 column level control	Maintenance planner
Replace C-101 Tray 7 sample valve	Process engineer
Replace broken gauges on C-101	Instrument supervisor
Order three C-101 Nutter trays	Project engineering
Decontaminate, open and clean shell pressure blast X-101 tube bundle, test and repair as needed	Operations
Fabricate new tube sheet for X-101	Project engineering
Decontaminate P-101A, send motor to vendor for recondition, set impeller clearance, set and align motor	Maintenance supervisor
Decontaminate P-101B, send motor to vendor for recondition, set rebuilt pump, set and align motor	Maintenance supervisor
Decontaminate X-102A and B, open and clean shell, blast tube and test, repair as needed	Operations
Decontaminate P-103A, send motor to vendor for recondition, set rebuilt pump, set and align motor	Maintenance supervisor
Decontaminate P-103B, send motor to vendor for recondition, set impeller clearance, set and align motor	Maintenance supervisor

equipment are much harder to conjure up, especially if there are long lead times involved.

A shutdown coordinator would be wise to develop a master material list of all items required for the shutdown. Indicate which items are available, on order, or currently held in the storeroom. The final storage location should be noted for all received items, as shown in Table 2-2.

A responsibility should be assigned to each item on the list. Be sure the storeroom will have the quantity you need. Also, ensure that the storeroom is staffed on the evenings and weekends during the shutdown. Simply having free access to the storeroom may not be enough. Stores personnel can use their vendor contacts to purchase emergency parts and materials. Set aside an area for parts and materials to be used during the shutdown. Tag or mark the parts with the shutdown number. Sometimes a locked area is necessary.

Table 2-2 Material and Equipment Responsibility

Item	Date	Order #	Delivery Date	Respon-sibility	Cost (or Est.)	Current Location
2 Lgths sched 40 SS pipe	5/1/94	23454	-	MTH	1500 est.	On order
Blower bearings 6320	N/A	N/A	N/A	GHN	580 act.	Storeroom
5 KW Portable generator	5/13/94	24673	7/12/94	TDJ	1000 est.	Vendor
Coal grate	5/12/94	23473	6/30/94	GHN	22,000 est.	Vendor

Persons with responsibility for the project should meet frequently. During these meetings the shutdown manager should ask questions about each potential job. It's best not to ask questions that elicit only a yes or no answer, such as "Will all the parts be here in time?" It's best to have the person responsible explain the situation rather than answer yes or no.

The following are some examples of good probing questions:

- What prefab work is required?
- What are the logical steps for this job?
- What are the manpower, material, equipment, tools, or engineering requirements for the job?
- When does the vendor think parts will arrive here?
- How much of this material is in the storeroom?
- What are the additional materials, tools, or equipment that have to be ordered for this job?

Shipping and Delivery

To ensure that all parts, material, and equipment are readily available, it is important that those items required from an outside source are delivered on time and in good working order. A good planner will be cognizant of requirements for delivery of large items, freight costs, demurrage and rental costs, the tracking of long-delivery items, and accounting for all these stocking needs.

Delivery of Large Items

The delivery of large items may require a crane to unload. These large deliveries should be timed so that the items can be unloaded and moved to their points of use. This will eliminate the need to bring the crane back a second time to move the item to where it is needed.

Saving on Freight Cost

Standard delivery terms are used in a contract to determine the point where the buyer takes possession of the goods and who is responsible for shipping costs. The buyer should specify the type of transportation to keep shipping costs down, or else the seller may use its own discretion. Often, the carrier provides a shipping receipt or bill of lading to identify the items that are being shipped. A *bill of lading* is a contract between the shipper and the carrier. The carrier must transport and deliver goods under the specified terms of the bill of lading.

F.O.B. (which stands for free on board) is the most common freight term. F.O.B. is usually followed by place of shipment, destination, vessel, truck, or other vehicle to identify where and how the customer will take possession of a good. The F.O.B. point is usually the location where title to the goods passes from the seller to the buyer. The seller is liable for transportation charges and the risks of loss or damage to the goods up to the point when the title passes to the buyer. The buyer is liable for such charges and risks after passing of title. Following are some common F.O.B. terms:

- *F.O.B. place of shipment*—The seller must ship the goods from the place identified and bear the expense and risk of putting them in the possession of the carrier. The seller is responsible for placing them in the hands of the shipper with proper documents. The buyer is responsible for the shipping costs, which will probably be included in the final invoice from the seller, and must make its own damaged shipment claims. *Ex-Factory* or *Ex-Works* are international terms that can replace F.O.B. place of shipment.

- *F.O.B. destination*—The seller must accept the expense and risk to transport the goods to the destination. Delivery of the goods must be at a reasonable hour. The buyer must provide a reasonable place to receive the items and must unload them at its own expense. The buyer may choose these terms to keep paperwork down because the carrier does not have to be paid separately by the buyer.

- *F.O.B. vessel, car, or other vehicle*—The seller is responsible for loading the goods on board a vessel, car, or other vehicle at the seller's own expense. F.O.B. vessel requires that the buyer identify the vessel and appropriate case. The seller must comply with overseas shipping law (if applicable), which includes obtaining a negotiable bill of lading. A buyer may choose this option because it has a favorable contract with a specific carrier. The buyer must pay the shipping and any insurance, which are usually billed separately by the carrier.

Other terms found on a purchase contract include the following:

• F.A.S. (which stands for free along side) is a pricing and delivery term used to name the port of entry for a product. The seller must (at the seller's own risk) deliver goods along side a vessel at a port or on a dock at the port. The buyer must designate the port. The carrier must present a bill of lading when the buyer presents a receipt for the goods.

• C.I.F. (which stands for cost, insurance, and freight) and C.&F. or C.F. (which stand for cost and freight) are terms used for marine transportation. These terms (often interchangeable) identify what the price includes. The seller must obtain a bill of lading covering all transportation to the destination (usually a port). The carrier must present the seller with a certificate of insurance. The buyer is responsible for transportation from the port to the end location.

• Ex-Ship means from the carrying vessel. Shipment is required from the carrying vessel to a normal dock or storage area and is not restricted to a particular vessel. As with F.A.S., the seller must obtain a bill of lading from the carrier.

Sellers often quote prices with a shipping stipulation of F.O.B. destination. This may not be the most advantageous terms for your company. The shipping costs (which are included in the price) are an average or nominal cost of shipping. In other words, if the supplier ships to a company 10 miles away or to one 2000 miles away, the shipping cost is the same. This may be acceptable if the company is 2000 miles from the supplier. However, for the company who is only 10 miles away, the shipping costs can be excessive.

Buyers ask for F.O.B. destination because it ensures that freight costs are applied to the products with which they are associated and minimizes the administrative effort involved in paying freight bills and collecting freight damage or loss claims. Unfortunately, it often does not result in the lowest freight cost to the buyer. Whenever a significant shipment of goods is to be made, the buyer should change the freight terms to F.O.B. place of shipment and arrange to route and directly pay for all freight costs.

Some companies acquire and further develop routing guides wherever possible. These guides are made up of lists of frequent company shipping points and carrier(s) that best serve the company from those points. With the development of routing guides and the management of freight costs by buyers, freight volumes can be determined and will serve as the basis for negotiations with carriers regarding rates and services (such as freight consolidation).

Demurrage and Rental Charges

The return of trailers, rail cars, gas cylinders, and any rented equipment should be included as an activity in the project schedule. The buyer may be responsible for all additional charges once the seller has met the delivery terms. If the terms of a contract allow goods to be delivered to a dock located away from the end destination, the buyer is responsible for all demurrage and additional transportation to the end destination. *Demurrage* is a charge assessed by a bonded warehouse, barge company, railroad, or trailer company for storage expenses beyond reasonable requirements. Demurrage can also be charged for gas cylinders during the excess time that the buyer holds the cylinders.

Tracking Long-Delivery Items

Obtaining a ship date from the supplier is usually sufficient for tracking such items for most short delivery items. If the supplier fails to meet the ship date, other sources are usually readily available. For long-delivery items (that is, items that are only available from a single source, or items that must be specially made), additional information may be needed to adequately track them. Obtaining only a ship date and then responding with expediting measures when the ship date is missed could spell disaster to the project schedule.

Some items with long delivery are only manufactured after an order is placed. The item is usually expensive, and a supplier is not willing to tie up cash in inventory. In any event, it is prudent to require the supplier of long-delivery items to provide a milestone calendar for manufacture and delivery. This calendar is a listing of how the order will progress from the time the order is placed until it arrives at the customer's site. As a minimum, the following dates should be obtained from the supplier:

- *Engineering and design assignment*—If the item on order is a specialty item requiring some engineering and design, the order will go into a schedule for this up-front work. Request the actual date that the order will be assigned for this to the supplier's engineering group.

- *Engineering and design completion*—This date will be the supplier's estimate of when the engineering and design work will be complete. This date is critical, as there may be an official sign-off on this work before fabrication can begin.

- *Fabrication or manufacturing assignment*—This will be the date that the supplier will begin fabrication. This date will follow the design completion (at a minimum) by whatever time is needed to order material for the job. Requiring the supplier to pre-order this material can minimize this time. Although some additional charge may be placed on the order for the early purchasing of this material, it may be a good idea.

- *Fabrication or manufacturing completion*—This is the projected date that the order will be completed. If the manufacturer does not operate around the clock, seven days per week, paying additional charges for premium rates can shorten time required for fabrication.

- *Quality assurance or performance testing*—Special order items often require some assurance or testing phase. The customer may want to be present for the test, called a *witness test*.

- *Preshipment packaging*—Large items usually require some form of preshipment packaging. The item may need to be temporarily bolted to cribbing or protected against climactic conditions in transit. Require the supplier to project how long this will take.

- *Ship date and shipping duration*—These dates are self-explanatory. However, they tend to slip. If shipment is scheduled by rail or sea, the projected dates can be affected by weather conditions.

When the above dates are provided, it is a good idea to follow up at or before each milestone. Make the supplier report the impact that a missed milestone has on future milestones. Insist that the supplier explain how the order will catch up with lost time if the supplier responds that there will be no change in subsequent dates. Staying on top of the order from the very beginning is the best insurance in keeping the order on track. Often, the supplier will give your order special attention in its processing just because of your persistence.

Purchase Order Expediting

Anyone who places an order is responsible for the follow-up for all rush, overdue, or back orders. This follow-up is called *expediting*. Following are some good disciplinary steps to accomplish this task:

1. Review your file for orders that are coming due. An *overdue order* or *back order* is any order that has not been completely received by the need, due, or promised date.

2. For each overdue order, check with receiving to verify that the material is still outstanding.

3. Call the supplier to ascertain revised shipping date. Inform the vendor if the revised shipping date is not acceptable.

4. Find another vendor for the product before canceling the order with the original vendor.

5. If the supplier states that the order has been shipped, get the shipping date, carrier, and freight bill number.

6. Call the carrier and initiate a trace.

7. If advised that material has been delivered, obtain name of the person who signed for receipt and the date received.

The time expended to expedite overdue and back-ordered materials is minimal compared to the problems caused if the material is not available for the shutdown.

Accounting

Good accounting is the key to cost control. Standard accounting practices still hold true during a shutdown. Be sure these procedures are not abandoned in the name of expediency.

It's a good idea to assign a number to the shutdown (for example, Shutdown #081295-003). This number should be marked on all purchase orders, work orders, stores slips, and contracts associated with the shutdown. If a question ever arises as to what a specific purchase is for, the shutdown number will answer that question.

It's also good accounting practice to reference a work order on all parts, materials, rentals, or contracts. Many plant accounting systems are already set up to tie purchase orders to separate temporary accounts, such as work orders.

Planning Thought Process

All jobs that become part of the total project should be carefully planned and estimated. The natural tendency when looking at a specific job is to identify the parts that might be required and then, as an afterthought, come up with an estimate. Unfortunately, if the job has not been completed in the past, the estimate is based on a guess.

One estimating method that uses the abilities of a planner to the fullest is the *planning thought process*. The planning thought process has three main purposes:

- Establish an estimate of time required for the job
- Define understandable steps to complete the job
- Identify the materials, parts, tools, and equipment required for the job

An approach that yields the best result is to visualize the job and how it will proceed. Visualization means forming a mental picture of a job and the persons working on it. This tends to bring more aspects of the job to light (such as the need for tools, parts, or equipment).

Steps developed through the planning thought process can be used by a maintenance worker as a guide to completion of the job, if this is required. Additionally, planning thought process estimates are developed in more manageable time units. Job step durations derived through the planning thought process vary from .25 to 4 hours.

The thought process begins by asking a set of questions about the job, as shown in Figure 2-3.

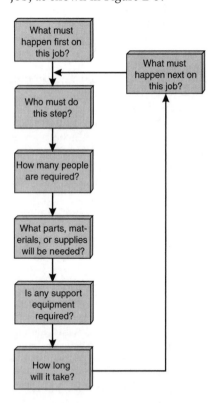

Figure 2-3 Planning Thought Process.

Identifying a step and then asking the questions about each step helps to develop a better overall plan. This thought process is described as follows:

1. *What must happen first on the job?* While visiting the job site, planners will normally ask themselves what must happen first

to begin the job. Does the operator have to be contacted? Will equipment have to be locked out? Will tools or parts have to be collected first? These are the common first steps of a maintenance job.

2. *Who must do this step?* A planner will next identify the crafts or minimum skill level required to complete this step. Skills required for one step might not be the same as skills required for subsequent steps. Sometimes an overqualified person may be allocated to a few steps in a job to better expedite the total job, rather than switching people in the middle of the job.

3. *How many people are required?* While thinking about the job step, the planner may determine that more than one person is required to complete it. The first inclination should be to determine the minimum number of people required. If reduced downtime is the goal of the plan, the planner may opt to staff this step with two or more people. The maximum number of people required on the step will be limited by the space available and physical conditions, all of which the planner can easily assess at the job site.

4. *What parts, materials, or supplies will be needed?* Many planners do a very good job determining the labor requirements but miss the material requirements. This is less likely to happen if the job is broken down into steps. As a step is reviewed, the parts, materials, and supplies required will be obvious.

5. *Is any support equipment required?* Just as with material, special tools or equipment required for a job are more easily defined when the job is broken down into steps. Is a fork truck required? Is a torque wrench required? Is a pick-up truck required?

6. *How long will it take?* Here is where the planners' experience is put to the test. Fortunately, when jobs are broken down in logical steps, elapsed time estimates for these steps will fall out naturally. The planner will usually multiply the number of people assigned to the job times the elapsed time estimate to determine the total labor hours.

7. *What must happen next on this job?* After all aspects of a specific step are reviewed the planner will identify subsequent steps and repeat the thought process to identify labor, material, and equipment requirements.

It's a good idea to write the steps down as they are visualized. A simple planning worksheet such as the one shown in Figure 2-4 is used as an aid to put the thought process down on paper.

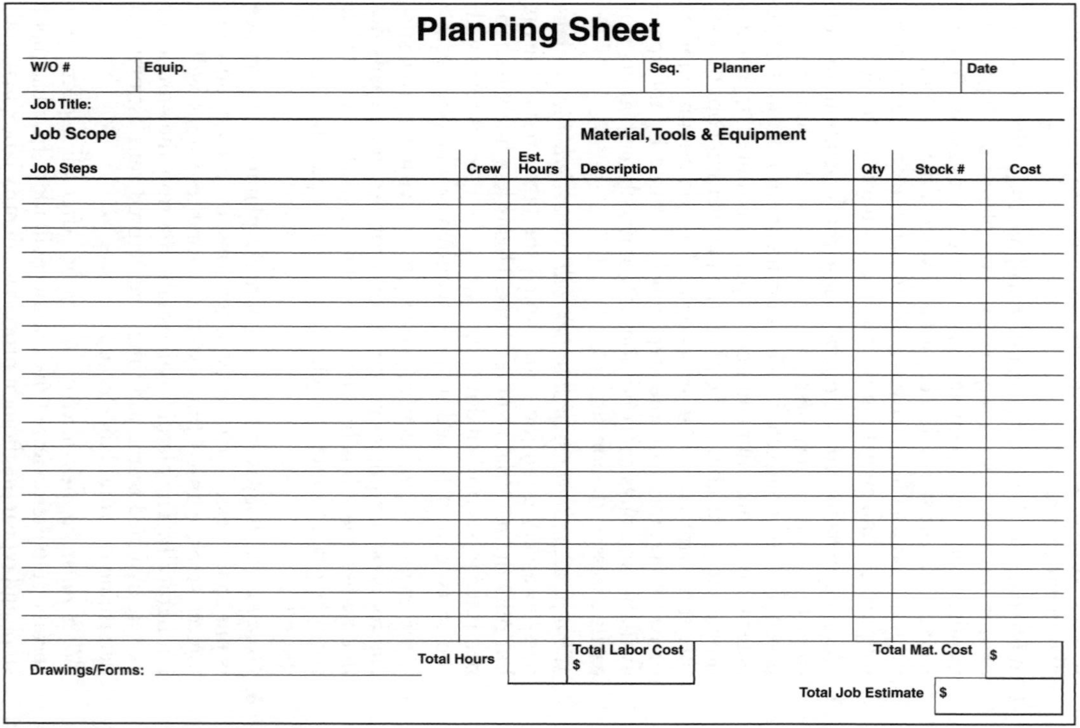

Planning Sheet

W/O #	Equip.	Seq.	Planner	Date

Job Title:

Job Scope			Material, Tools & Equipment			
Job Steps	Crew	Est. Hours	Description	Qty	Stock #	Cost

Drawings/Forms: ______ Total Hours | Total Labor Cost $ | Total Mat. Cost $

Total Job Estimate $

Figure 2-4 Planning Sheet.

This planning sheet is designed to step the planner through the thought process. It's filled out in the following manner:

1. The *W/O#* (work order number) and *Equip.* # (building or equipment identification) are used to tie the planning sheet to a specific job.

2. The *Seq.* (sequence) is used if multiple crafts exist. A planning sheet is filled out for each craft to be used and the sequence (some work order systems refer to this as a job step) ties all crafts to the original work order.

3. The *Job Scope* is where the thought process is noted. It is useful to write down each specific step and estimate the labor hours needed to do that step. Some planners tend to estimate job labor hours in multiples of four hours. In other words, without rigorous estimating, job estimates tend to be 4, 8, 16, or 32 hours in duration. This is sloppy estimating that removes the ability to truly optimize resources during the scheduling process.

4. The *Material* section allows for identifying any parts or materials needed. As the job scope is being developed and needed parts come to mind, it is best to note them immediately.

5. Any special *Tools and Equipment* are noted. This could entail special hand or power tools or any other items not normally included in a tool cart or pouch. Support equipment, either in-house or contracted, is also noted here.

6. Any *Drawings/Forms* can be referenced. Many planners will use the back side of the planning sheet for a simple sketch and incorporate the planning sheet itself as part of the work order.

When the planning sheet is complete, it can be used as an input document for a CMMS or can just be filed for future reference.

A well thought out planning sheet can be used to develop a plan for most job types. The following is an example of how this sheet can be used.

Example of a Turnaround Planning Sheet

A typical turnaround in many process plants involves the cleaning of heat exchanger tube bundles. The planning sheet shown in Figure 2-5 was developed for such a job.

Planning Sheet

W/O #	Equip.			Seq.	Planner			Date
49501-2	X-102A Amine Cooler - North				TMH			3/22/01

Job Title: Annual Turnaround Cleaning & Inspection

Job Scope

Material, Tools & Equipment

Job Steps	Crew	Est. Hours	Description	Qty	Stock #	Cost
1. Close, lock & tag water inlet and outlet valves, pull elbows and blind amine inlet and outlet valves.	2 PF	.5	2 valve locks, 2 8"-150# blinds			
2. Unbolt head, rig and move to end of monorail. Set on cribbing.		.75	1/2 ton chain fall, 10' x 3/4" cable choker; 4 x 4 cribbing			
3. Attach eyebolt to tube sheet, rig to monorail. Use 2 ton chain fall to pull bundle. Cradle end of bundle with nylon strap and chain fall.		.5	2 ton chain fall, 1" eyebolt, 3" x 10' 2000# nylon strap, 1/2 ton chain fall			
4. Set up curtains and pressure blast tube bundle and shell. Do not exceed 10,000 psi water pressure. Clean gasket of water and amine side of tube sheet.	CTR	3.0	HydroClean, Inc.			750.00
5. Inspect tube bundle and shell.	1 Insp	1.0				
6. Install new gasket on water side of bundle. Reinstall bundle into shell.	2 PF	1.0	32.25 x 34 x 1/8 Garlock 3300	1 ea	34-510-132	54.95
7. Install new amine gasket. Lift amine head and install on shell. Use new 7/8" nuts.		.75	150#, 32" Grafoil spiral wound 7/8" Grade 2H nuts	1 ea 48	34-275-532 14-278-010	592.00 free issue
8. Lubricate and torque head studs to 270 ft-lbs. Progress in 90 ft-lb increments. Use criss-cross pattern in tightening sequence.		.5	0-500 ft-lb torque wrench (tool crib)			
9. Reinstall elbows. Remove locks and tags.		.25	8"-150# Grafoil spiral wound	4	34-275-008	89.80

	Total Hours	9.5	**Total Labor Cost** $ 351.50		**Total Mat. Cost**	$ 1486.75

Drawings/Forms:

Total Job Estimate $ 1838.25

Figure 2-5 Planning Sheet for a Turnaround Job.

The following plan details the work necessary to clean and inspect an amine cooler (refer to Figure 2-5). The estimate was developed in nine steps, as follows:

1. The first step is to close and tag the cooling water valves and to remove an elbow and blind the block valve on both the inlet and outlet amine lines. Although very little activity is involved in these actions, a .5-hour step duration is estimated as this first step also includes time to assemble all necessary tools and materials before proceeding to the job site. Two valve lock assemblies and two 8-inch blinds are identified as needed materials. Two pipefitters will be assigned to this job.

2. Next the head must be supported with a small chain fall and then unbolted and moved down an installed monorail provided for this repetitive job. Because the chain fall will need to be reused later, the head will be lowered to cribbing material during the cleaning and inspection operations. A .75 hour duration is estimated for this step and in addition to the chain fall unit, additional material and tool needs include a $3/4$-inch by 10-foot cable choker and the four-by-four cribbing.

3. After the head is removed, a 1-inch eyebolt is attached to a tapped hole in the center of the tube sheet. The original chain fall is reused to support the tube sheet from the monorail, and an additional larger chain fall is needed to pull the bundle free from the shell. The opposite end of the bundle must be cradled with a nylon strap and supported with a second small chain fall attached to the monorail in order to pull the entire bundle free from the shell. It is estimated that this step will take .5 hours.

4. An outside contracted firm will do the actual cleaning of the tube bundle. This step is included in the plan because the maintenance supervisor will also oversee the contractor's work. The planner has noted in this step the need to limit the pressure the contractor should use during the cleaning and the requirement to also clean the gasket surfaces on both sides of the tube sheet. The three-hour duration of this step has been discussed and agreed to by the contractor during an earlier meeting.

5. After the cleaning is complete, the bundle must be inspected. Someone will do this from engineering. The one-hour estimate is given so that the supervisor can anticipate when the pipefitters need to be reassigned to the job to complete it.

6. Before installing the tube bundle, a new gasket must be installed on the water side of the tube sheet. One hour is

estimated as the time required to slide the bundle back into the shell. The proper gasket is noted in the material description as well as the storeroom part number because this gasket will have to be requisitioned at the store's counter.

7. Before installing the head, a new gasket is also needed for the amine side of the tube sheet. Again, the proper gasket is identified as well as the storeroom part number. Since the old nuts should not be reused, the planner noted that 48 new $7/8$-inch Grade 2H nuts needed to be retrieved from free issue. The storeroom part number for these nuts is identified in case the free issue stock is insufficient. This step is estimated at .75 hours.

8. The head studs need to be torqued for proper assembly. A 0–500 ft-lb torque wrench will be needed from the tool crib in order to achieve the necessary torque value. It is estimated that this step will take .5 hours.

9. The final step is to reinstall the amine elbows and remove the locks from the cooling water valves. A quarter hour is estimated for this last step. Four gaskets are also identified as needed in reinstalling the elbows.

The total job is estimated at a cost of $1838.25. This estimate will be needed in developing an estimate of the overall turnaround cost before the turnaround actually begins.

The actual work order itself does not necessarily require the detail included in the planning sheet. The planner developed this detail to ensure that the full scope of the job was being identified, as well as a complete listing of all resources needed beyond the labor resource itself. The actual work order might appear as shown in Figure 2-6.

The work order itself provides a brief description of the work to be performed. The planner is aware that the pipefitters and their supervisor know what is needed to do the job but indicates the proper torque value to be achieved. In addition, the need to perform the torquing in increments as well as using a criss-cross sequence when tightening the nuts is noted. All needed materials and tools are identified so that unnecessary trips from the job back to the maintenance shop or storeroom should be eliminated. The stock number for the $7/8$-inch nuts was included in case there are not enough available in the free issue area of the storeroom.

Applying the Planning Thought Process to Large Jobs
The planning thought process helps identify and understand what is needed to complete a job, but it can also be applied to each phase of

Maintenance Work Order						#49501-2

Date Initiated	Originator	Downtime?	Date Available	Z		Priority
03/17/01	SMITH	YES - Amine Plant	07/14/01		08:00	PLANNED

Equip. #	Equipment Description		Date Required
X-102A	AMINE COOLER		07/14/01

Classification: PREVENTIVE

Work Requested: ANNUAL T/A CLEANING & INSPECTION

LOCK & TAG COOLING WATER VALVES, PULL AMINE ELBOWS & BLIND
VALVES. REMOVE HEAD, PULL BUNDLE AND INFORM HYDROBLAST
CONTRACTOR AND INSPECTION. AFTER INSPECTION, REINSTALL
BUNDLE USING NEW GASKETS. USING NEW NUTS, TORQUE HEAD
TO 270 FT-LBS IN 90 FT-LB INCREMENTS. USE CRISS-CROSS
TIGHTENING SEQUENCE.

Material & Parts Description	Stock #
32.25 X 34 X 1/8 Garlock 3300	34-510-132
150#, 32″ Garfoil spiral wound ring gasket	34-275-532
150#, 8″ Garfoil spirol wound ring gasket - 4 reqd	34-275-008

Tools: 2-1/2 Ton Chain Falls; 1-2 Ton Chain Fall, 1-3/4″ x 10′ cable
chocker, 3″x 10′ 2000# nylon strap; 1-1″ eye bolt; 48 - Grade 2H
nuts (14-278-010); 0-500ft-lb torque wrench; 4 x 4 cribbing

Safety Checks: LOCK AND TAG BEFORE BEGINNING WORK

Comments:

Figure 2-6 Job Work Order.

a larger job. Many planners have made the observation that a job to
be performed on production, manufacturing, or process equipment
will proceed through the following phases:

- Assignment and procurement
- Travel and staging
- Equipment clearance and entry
- Scope of work
- Return to service
- Clean-up and travel
- Closure and documentation

Assignment and Procurement

The assignment and procurement phase is the time required for supervision (management) to relay job information to the maintenance

personnel who will perform the work and the time workers will need to collect and assemble resources. This time might be relatively short and almost inconsequential if the job is straightforward or relatively simple in scope. However, the time required for this phase could constitute an hour or more if the job is complicated. There could be special safety considerations or extensive procedures governing the work. This phase may require a site visit by the planner to discuss job particulars on some jobs. In any event, the time required to relay job information and assemble resources must be accounted for in the job estimate.

Travel and Staging
Travel and staging recognizes the time needed to locate personnel and resources to a job site. Many planners develop guidelines to quickly estimate the travel time to get to specific plant areas or elevations. Additional resources that need to be recognized for this phase include mobile equipment and rigging that may be needed to move spare parts or equipment. In facilities covering many acres or square miles of property, travel and staging can be significant and is often referred to as *windshield time*. In smaller facilities, the travel and staging aspect of a job is often combined with the first phase of assignment and procurement.

The travel and staging phase must also identify what utilities will be required at the job site and what resources need to be staged to provide these services. Extension cords and multiple outlet boxes may be required to provide 120-volt service for power tools or lighting. Additional hoses may be needed to bring service air to the job from the nearest connection.

Equipment and Clearance Entry
The equipment clearance and entry phase includes the activities that must take place to prepare the equipment, the site, and the workers for work. Some of these activities may be the responsibility of operations or production personnel and do not represent time that maintenance workers will charge to a work order. Other activities may require the presence of both maintenance and operations to ensure the safety of the workers. These activities must be recognized and addressed in the planning effort and included in the scheduling effort. The following three specific areas must be addressed in equipment clearance and access:

- *Equipment or Machinery Preparation*—The equipment or machinery is prepared for opening or physical entry by maintenance workers. It may require purging or some other decontamination step to clear the equipment of hazardous or volatile components. It will probably also require some type of test to ensure that the hazardous or volatile material is absent. In the case of physical entry, oxygen content must be verified or self-contained breathing apparatus must be specified for all workers. This equipment preparation activity is usually the responsibility of production or operations personnel. However, the planning estimate should include time for maintenance workers to witness tests or inspect testing results.

- *Job Preparation*—This activity includes specific steps that must be taken to prepare for the actual work. Since maintenance personnel will generally perform these steps, they must be included in the work estimate or may be accounted for through other work orders. (If separate work orders are written to assign to other crafts, these jobs should be tied to the parent work order in order to provide total accounting once the job is complete.) Job preparation might include erection of scaffolding or procurement of ladders. Insulation may have to be removed before the principal workers can be assigned. Mobile equipment (such as man-lifts) should be identified in this phase. Job preparation might also include specific safety procedures (such as tagging, locking, and blinding of machinery or equipment). The planning effort should attempt to quantify this information so that unnecessary trips aren't required to pull these resources together. It might be necessary to include sketches or drawings to stipulate exactly where tags, locks, or blinds are to be located so there is no chance that work can start on machinery or equipment that is improperly isolated.

- *Worker Preparation*—This activity identifies specific steps that must be taken to prepare the workers before work can begin. It will include Personal Protective Equipment (PPE) that must worn or used, depending on the hazards involved or the final cleanliness of the equipment being repaired. The time required to collect and suit up in protective gear must be included in the work estimate.

Equipment clearance and entry on a small job that requires no permitting may be combined with the first two phases as a single job step. On a large job, with extensive decommissioning of equipment, this phase may require several hours of time just to properly tag, lock, blind, test, and verify that procedures have been completely followed before work can begin.

Scope of Work

The Scope of work is the actual maintenance activity. It is best defined using the planning thought process, as this method ensures that the work is broken down into understandable job steps and each step is thoroughly analyzed to identify all resources. On particularly large or complicated jobs, it is helpful to further break the scope of work down into the following four broad steps:

- *Disassembly*—This entails all steps necessary to tear equipment or machinery down to the point of repair or inspection. These necessary steps are often detailed in equipment manuals, which can be substituted for job steps or procedural steps.

- *Inspection or Repair*—This work covers all clearances or tolerances that have to be checked or verified. It can also include specific components that must be replaced. Again, equipment manuals are an important source of information and detail.

- *Reassembly*—These steps are basically a reversal of the disassembly procedures. What must be included in this work, however, are torque values, tightening sequences, and replacement of expendable supplies such as gaskets, seals, oils, or coolants.

- *Testing*—This work normally covers any precommissioning or testing prior to releasing the equipment. It may, however, involve performance testing that can't be conducted until the equipment has been released and brought back into service.

Return to Service

Return to service involves removal of any tags, locks, or blinds. The planner should identify replacement gaskets, nuts, or other expendables. This phase should also include time and resources for insulating and tearing down scaffolding, even though this work might not be performed until later.

Clean-Up and Travel

The clean-up and travel phase covers any time required to dispose of used parts or equipment removed during the scope of work. It also includes disposal of any contaminated supplies or solvents generated by the work. Finally, it also includes time spent in returning mobile equipment or ladders and travel time of the workers to a base location.

Closure and Documentation

The closure and documentation phase covers the time needed to add detailed information to the work order before it is closed out. Such details include actual hours charged to the work order, additional parts, materials, or supplies that were not specifically charged against the work, and as-left details or readings (clearances, alignment status, and so on). The trend in industry is to enlist the workers in providing this information as part of the total job.

On a relatively small job, the posting of actual work hours to a work order may be all that is needed to complete the record for history (assuming that material resources were already charged to the work order when these items were received). On a more complicated job, such as the overhaul of a large piece of process machinery, closure and documentation might require a post-work interview with the workers to ensure that all pertinent information was collected and committed to the historical record.

On critical work, such as major overhauls performed during an outage or shutdown, specific time must be allocated in the job estimate so that this important part of the job is not overlooked or dealt with in a cursory manner.

Example of Planning a Large Job

The power assembly (bearing housing, casing adapter, shaft, and impeller) of an 8 × 10–15 horizontal centrifugal pump (P-101A) needs to be changed in the field. The pump is an ANSI back pullout design, and the plant utilizes spacer dropout couplings so that the motor driver does not have to be disconnected and moved. The pump operates at 250°F and the material being pumped causes moderate to severe chemical burns unless it is diluted with copious amounts of water.

Plant safety procedures require that a minimum of two maintenance personnel suit up with rubber suits, rubber gloves, and face shields when initially opening process connections. Additionally, the inlet and outlet valves must be locked in the closed position and skillet blinds must be installed at the suction and discharge flanges of the pump casing before work can proceed.

The following plan indicates the procedural steps, resources, and step durations for each phase of the job. The pump and driver are located at ground level approximately 200 yards from the maintenance shop. The pump driver is a 150 HP, 1750 RPM, 460 volt, 445T frame AC induction motor. The job plan is shown in Figure 2-7.

The planning sheet developed the job scope into nine steps. The job phases were incorporated in the steps as follows.

- The assignment and procurement phase and the travel and staging phase are incorporated into Step 1. Two mechanics

Planning Sheet

W/O #	Equip.		Seq.	Planner		Date
100269	P101A			TMH		12/15/01

Job Title: Change Power Assembly

Job Scope

Job Steps	Crew	Est. Hours
1. Stage spare unit, tools, and materials to jobsite.	2 ME	.5
2. Verify pump clearance from OPS, lock/tag same.		
Follow Acid Line Entry Permit on casing flanges.		.5
3. Rig pump at frame adapter. Drop spacer coupling,		.75
remove casing and bearing frame bolts, and jack out		
pump. Wash pump and casing to chem sewer.		
4. Install pump coupling hub on new assembly. Set new		1.0
unit on base. With new gasket, install assembly into		
casing. Torque casing bolts to 30 ft-lbs. Set		
impeller to casing clearance at .017".		
5. Install spacer coupling. Use new coupling insert.		.25
6. Align motor to pump. Target motor shaft .005"		1.5
high to pump shaft. Record final readings.		
7. Remove skillet blinds. Replace casing flange gaskets.		.75
Torque 10" flange bolts to 270 ft-lbs and 8" flange		
bolts to 180 ft-lbs. Follow Acid Line Entry procedures.		
8. Remove all locks. Inform OPS and witness run-in.		.5
Record overall vibration levels.		
9. Deliver old pump to Pump Shop. Close out job.		.75

	Total Hours	**13.0**

Material, Tools & Equipment

Description	Qty	Stock #	Cost
8x10-15 power assembly	1 ea	85-210-000	8175.00
Drott 85 Carry Deck			
2 valve locks, 8" & 10" skillet blinds			
6' cable choker			
casing gasket		85-210-351	104.95
0-150 torque wrench		(tool crib)	
dial indicator w/ mag base		(tool crib)	
Size 5 Hypalon coupling insert	1 ea	12-275-532	54.00
laser alignment unit		(tool crib)	
"C" shim kit		(tool crib)	
8"-150# Grafoil spiral wound	1 ea	34-275-008	73.04
10"-150# Grafoil spiral wound	1 ea	34-275-010	92.17
0-500 ft-lb torque wrench		(tool crib)	
hand held vibration meter		(tool crib)	

Total Labor Cost	**Total Mat. Cost**
$ 533.00	$ 8499.16

	Total Job Estimate	$ 9032.16

Drawings/Forms: Alignment Data Sheet
Vibration Data Sheet

Figure 2-7 Large Job Plan.

will be assigned to the entire job. The resources needed include the new power assembly for the pump and the Drott 85 Carry Deck. A duration of 0.5 hours is estimated for this step.

- The equipment clearance and access phase is covered in Step 2. Operations will have been informed about the scheduling of this job the day before, and isolation of the equipment is supposed to occur on the graveyard shift. The mechanics will need to verify this has been done. The reference to the Acid Line Entry procedure will inform the mechanics about the personnel protective equipment (PPE) that must be brought to the job (that is, rubber suit, rubber gloves, and face-shield). The valve locks and skillet blinds are identified as necessary resources and a 0.5-hour duration is given to the step.

- The scope of work phase is covered in job Steps 3 through 6.

  Step 3 involves disassembly of the pump and decontamination of the pump unit. A 6-foot cable choker is identified and 0.75 hours are allowed, as the mechanics have to perform this work still wearing the PPE.

  Step 4 covers installation of the new unit. The equipment manual was consulted to obtain the impeller-to-casing clearance setting, as well as the final bolt torque required. A new casing gasket will be needed. A dial indicator with magnetic base will be needed to set the impeller clearance and a 0–150 ft-lb torque wrench is required as well. A one-hour duration is estimated for this step.

  Step 5 is mainly a reminder to replace the coupling insert.

  Step 6 covers the final alignment. The cold alignment target was recommended by the equipment manual based on the 250°F operating temperature. A duration of 1.5 hours is allowed, and the laser alignment unit and "C" shim kit is also identified.

- The return to service phase is covered in Step 7. The different torque levels for the suction and discharge flange bolts are based on the maximum allowable assembly torque for Grade 5 bolts, the company standard. The need for two new gaskets and a 0–500 ft-lb torque wrench are noted. A duration of .75 hours is allowed for this step, as the mechanics have to get back into the PPE for this work.

- Step 8 also involves some of the return to service phase, since it includes removal of any final locks and tags. It also incorporates a test procedure because the mechanics must witness the

> start-up of this pump and obtain and record some vibration readings. About .5 hours are allowed because the operations department may not be able to start up the unit as soon as the return to service is complete.
>
> • Both the clean-up and travel phase and the closure and documentation phase are included in Step 9, and .75 hours are estimated as the duration of this last step.

The labor estimate of 13.0 hours can be translated into a cost estimate of $533.00 (based on a base labor rate of $41/hour). The material costs are estimated at $8499.16 for a total job cost estimate of $9032.16.

The work order that results from this plan does not have to reflect all the details of the planning effort. It is important to show the impeller clearance, cold alignment target, and flange bolt torque on the casing bolts. The real value of the planning effort is evident in the identification of the materials required for the job. These parts can be put together in a kit, delivered to a shop, or staged at the job site before the job is scheduled. Finally, identifying special tools and mobile equipment will help this job go as planned. A sample work order is shown in Figure 2-8.

The effort that went into planning this job does not have to be duplicated for every pump power assembly job in the future. In fact, the plan represents a starting point for any other pump job in the plant. All that needs to be modified are the specific resources that might be needed. Also, some of the torque or clearance values will change for different size equipment or different operating conditions. However, the fact that this plan identifies those values in this job would be a reminder to a planner that those values must be verified for a different job. Planning future pump replacement jobs would be a much simpler task because of the effort put into this first one.

Estimating Tips

Many jobs can be broken into piecework functions and the time required to perform each piece can be timed or estimated. The resulting per-unit time can be multiplied by the number of times the unit must be performed to determine the time required to perform the whole task. This estimating tool is referred to as *scaling* because the per-unit value is used to scale a job estimate based on the number of repetitions of a given activity. PM tasks often fall into this category.

For example, assume a greasing route includes 45 motors and it takes 10 minutes to properly grease each motor. The greasing

Maintenance Work Order				100269		

Maintenance Work Order   100269

Date Initiated	Originator	Downtime?	Date Available	Time Available	Priority
12/13/01	SMITH	No (running spare)			PLANNED

Equip. #	Equipment Description	Date Required
P101A	NORTH CIRCULATING PUMP	

Classification: CORRECTIVE

Work Requested: Change Power Assembly
Stage parts and set up temp water. Isolate pump per Lockout/Tagout
and Acid Line Entry procedures. Rig pump at Frame Adapter. Remove
and wash pump and casing to chem sewer before removing PPE. Set new
pump and torque casing bolts to 30 ft-lbs. Set impeller to casing
clearance at .017". Install new coupling insert and align pump
shaft .005" low to motor. Suit up in PPE before pulling blinds.
Install new flange gaskets. Torque suction flange to 270 ft-lbs and
discharge flange to 180 ft-lbs. Remove locks/tags. Inform OPS and
witness run. Record vibration readings. Deliver old pump to Pump
Shop and close out job in CMMS. Scan Data sheets into history record.

Material & Parts Description	Stock #
8x10-15 power assembly	85-210-000
8x10-15 casing gasket	85-210-351
Size 5 Hypalon coupling insert	12-275-532
150#, 8" Grafoil spiral wound ring gasket	34-275-008
150#, 10" Grafoil spiral wound ring gasket	34-275-010

Tools: (Alignment and Vibration Data Sheets attached)
Drott 85 Carry Deck; 2 valve locks; 8" & 10" skillet blinds; 6'
cable choker; 0-150 & 0-500 torque wrenches; dial indicator w/
mag base; laser alignment unit; "C" shim kit; vibration meter.

Safety Checks: LOCK/TAG & ACID LINE ENTRY PROCEDURES IN FORCE
Set up hose with running water before suiting up to break flanges.

Comments:

Figure 2-8 Large Job Work Order.

portion of the job should take 7.5 hours (45 motors × 10 min./motor, 60 min./hr.). An estimate of setup time, travel time, and cleanup time can be added to 7.5 to find the total time for the job.

Scaling can take another form. Some tasks take longer or shorter periods to perform based on the dimensions of the object on which the work is performed. One portion of a painting job is a good example of this. It may take 18 hours (not including preparation and cleanup) to apply one coat of paint to an open cylindrical tank with a brush. If the tank is 24 feet in diameter and 15 feet high, then 1130 square feet (area = 3.14 × 24 × 15 = 1130) can be painted in 18 hours. It can then be computed that it takes .016 hours

(18 divided by 1130 = .016) to paint one square foot. If another tank surface to be painted were 4000 square feet, it would probably take 64 hours (4000 × .016 = 64) to complete the painting portion of the job.

It is important to note that scaling factors should be determined from actual activity. A planner should closely monitor jobs that can yield good scaling factors for future use. Additionally, estimates will still have to be made for the time required for setup, cleanup, travel, or lockout.

It is also important for planners to have a resource book with some general geometric formulas handy for calculating such things as the circumference of a circle, area of a circle, area of a triangle, volume of a cylinder, as well as other calculations.

Determining Elapsed Time and Job Staffing

Development of an efficient project plan requires that the elapsed time be established for each job. The elapsed time required for many jobs depends on the number of people assigned. A planner should review the requirements of the job and decide how many people can be safely assigned to a job. The goal would be to assign the most people who can perform the job at maximum efficiency.

The planner should review all jobs that can be performed concurrently after a shutdown schedule has been developed. It may be practical to reduce the number of people assigned to some jobs if resources are limited.

Care should be taken when changing the number of people assigned to a job in a project management computer program. Some project programs change the elapsed time automatically when you adjust the resources. For example, assume you have determined a job can be completed in 40 hours of elapsed time with one person. If you later feel this job requires two people to complete, then the computer program will automatically change the elapsed time of the job to 20 hours. The assumption made by the program is that both people will be working concurrently. This may not be the case. From time to time, one person may have to wait for the other to finish a task before both can rejoin the activity.

Table 2-3 is an example of how inefficiencies can be tolerated, to a point, if downtime must be limited.

The planner estimates 16 labor hours for this job. If two pipefitters are assigned, the job should take 8 hours (16 hours divided by 2 pipefitters), resulting in an actual 8 hours of downtime for the job. In this first case, the elapsed ideal equals the actual elapsed time because the two pipefitters can work independently and out of each

Table 2-3 Job to Replace an Acid Line

Est. Labor Hrs.	Crew Size	Elapsed Ideal (Est./Crew)	Elapsed Actual	Actual Labor Charged
16 Labor Hrs.	2 PF	8 Hrs.	8 Hrs.	16 Labor Hrs.
16 Labor Hrs.	4 PF	4 Hrs.	4 Hrs.	16 Labor Hrs.
16 Labor Hrs.	6 PF	2.7 Hrs.	3 Hrs.	18 Labor Hrs.
16 Labor Hrs.	8 PF	2 Hrs.	3 Hrs.	24 Labor Hrs.

other's way. The labor hours charged to the job should equal the estimate.

In an attempt to shorten downtime on the line, more people can be added to the job. Downtime is cut in half if four pipefitters are used on this job because the ideal elapsed time again equals actual elapsed time. Only 16 labor hours are charged to the job.

Downtime should be cut to 2.7 hours if six pipefitters are on the job. Unfortunately, inefficiencies creep into the job because of limited space and redundant activity. The actual elapsed time is really 3 hours. The labor hours charged to the job have increased to 18 hours because of this inefficiency.

When eight pipefitters are assigned to the job, it still takes 3 hours to perform. Twenty-four labor hours are charged to the job. The extra pipefitters provide no increase in efficiency and no decrease in downtime. Six pipefitters should be assigned to the job if low downtime is important. Otherwise, two pipefitters will suffice.

The inefficiency associated with more than one person on the job may be tolerated if equipment downtime must be kept to a minimum. However, there is an obvious upper limit to the number of people who can be assigned to a job and still reduce the downtime the job creates.

Including Time Required by Operations
Many project plans go awry during the execution phase because the planner forgot to take the needs of the operations department into account. The operations department usually requires some time to shut down, decontaminate, or lock out equipment in the affected area. As a result, precious project time is expended for these activities.

The shutdown coordinator should solicit data on the elapsed time required to perform preparatory tasks. An example of a list may look like the information in Table 2-4.

These activities and estimates should be added to the project schedule to improve the overall performance.

Table 2-4   List and Estimate of Operations Activities

Activity	Estimate
Lock out BL305 dryer blower	.5 hours
Cool down FR312 boiler	16 hours
Purge T314 tank	24 hours
Deactivate T314 piping	8 hours
Clean S317 screen unit	4 hours
Empty T310 and clean mixer	6 hours

Work Packages

Many jobs performed during day-to-day activities are carried out without the need for specific step-by-step instructions or detailed material lists. However, activities during a shutdown may be performed infrequently. This work may require detailed information to make sure nothing is missed.

Planners should collect all documents required to perform a job into a work package. A work package is a folder or group of folders, which includes all the necessary paper work and reference material required to complete a job. An example of one such package is shown in Figure 2-9.

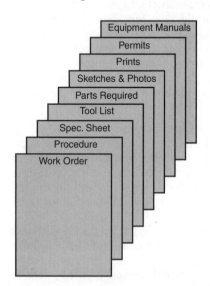

Figure 2-9   Work Package.

Note the following:

- A copy of the work order should be included along with a written procedure of how the job should go.
- A specification sheet should be added for jobs that must meet plant standards. Specification sheets (which indicate the goal for the repair process) should also include some space for a mechanic to report as-found and as-left readings. An example of this would be an alignment spec sheet to show dial indicator readings, before and after.
- Tool and parts withdrawal slips should be filled out ahead of time and included in the package.
- The planner should include sketches and photos of the work to be performed.
- Prints of the equipment or job site should also be included.
- All permits required to perform the work (such as line entry permits and lock-out tags) should be included.
- Equipment manuals can also be enclosed to help the maintenance worker through some tough procedures.

One use of a work package is to provide detailed information for contractors. A well-written work package can be handed over to a contactor and can be completed without much supervision by facility personnel. However, it is important to have a good written contract for services to ensure that the facility gets the best price for the work performed.

Contracts for Services

A contract should be drafted whenever construction or other services are required. The form of the contract can be informal (such as a purchase order referencing a proposal by the contractor). It can also be very formal, requiring bids along with detailed descriptions of the work and expectations of how the work will progress.

Types of Contracts

Contracts for services take a number of forms. Lump-sum and unit-price contracts are commonly used as a result of bids submitted by more than one contractor. Negotiated contracts can be let by a company to one or more firms. The contractor may be chosen, not so much for price, but rather for dependability, experience, or skill. The payment method can be lump-sum, unit-price, combination lump-sum and unit-price, cost reimbursable with a ceiling, cost plus a fixed fee, cost plus a percent of costs, construction management contracts, and incentive-type contracts. Following is a description of various types of contracts:

- *Lump-sum*—A lump-sum contract (also called *stipulated sum*) is suggested whenever it is possible for a potential contractor to make a precise survey of the work involved or when detailed drawings are available. The contractor will be paid one price (a lump sum) for the work performed, independent of the costs and time expended.

- *Unit-price*—When it is not possible to identify the battery limits of a project, or drawings are not sufficient, a unit-price contract may be preferred. With a unit-price contract, the contractor must estimate the labor and material used in definable units, (such as number of hours or number of cubic yards of concrete required for the project). These units are then multiplied by a unit price to obtain the extended cost for each item. Payment to the contractor will based on the actual number of units expended, as opposed to the estimated units.

- *Lump-sum and unit-price*—This contract type is used when some parts of a potential job are well documented but others are not. A lump-sum bid is submitted for the known portion, and a unit-price bid is submitted for the less-known portion.

- *Cost reimbursable with a ceiling price*—In this agreement, the contractor is to be reimbursed for all costs as listed in the agreement up to a maximum cost. The words *not to exceed* may appear on a purchase order to affirm the maximum cost.

- *Cost plus fixed fee*—This is an agreement similar to cost reimbursable with a ceiling except that a fixed fee is established for services rendered. This fee must be paid in addition to the itemized reimbursable costs. The terms of the contractor must define the points where completion will be measured and controls must be put on certain expenditures.

- *Cost plus percentage*—This contract is the same as cost plus fixed fee except that a percentage of the reimbursable cost is added to obtain the final payment. This may be undesirable to some buyers because the contractor's compensation for services increases with the costs of the project. The contractor's incentive may be to build up costs to increase its fee.

- *Construction management*—This type of contract requires the contractor to divide the work into trade segments. The contractor is usually the prime contractor who acquires bids from and hires trade subcontractors. The buyer reimburses the prime contractor for all subcontracts, as well as a profit for the prime contractor for managing the contract.

- *Incentive-type*—Incentive-type contracts provide for bonuses to be paid to a contractor for completing the job ahead of time or penalties for late completion.

Contract Revisions
Contracts or bids must be changed from time to time. Certain standard documents are required to explain these changes:

- *Addenda*—Sometimes revisions must be made during a bidding process. Addenda are provided to clarify or add drawings and specifications. These changes are usually a result of questions posed by prospective bidders. Addenda can also be used to change the dates of the bidding process.

- *Stipulation*—The buyer may modify the contract after the bid is awarded but just before execution. This change is provided in the form of a stipulation document. The contractor must agree to this stipulation prior to beginning work.

- *Change Order*—Changes in a contract while work is proceeding must be documented with a written order provided by the buyer. More often than not, a change order is applied to changes in a lump-sum contract. A change order is used when minor changes have occurred and when the general scope of the work has not. Changes that exceed a certain percentage (such as 25 percent) may be too excessive for a change order. In these cases a supplementary agreement must be prepared.

- *Supplementary Agreement*—When changes occur that are outside the general scope of the original contract, a supplementary agreement must be developed. In some cases a supplementary agreement can be a complete rewrite of the contract. Both parties must sign this new agreement.

The Bidding Process
Bids are often required for construction work and services when many different contractors can perform a particular job. Companies use the bidding process as a cost control measure and as a method to reduce collusion.

The bidding process begins with an advertisement or invitation for bids. Sealed bids may be invited by advertising in newspapers and engineering publications for legally required periods. The advertisement should contain an issuing location, date of issue, date for receipt of bids and time of opening of bids, brief description of the project, location of the project, quantities of major items of work, office where plans and specifications can be reviewed, proposal security, and rights reserved to the owner.

More often than not, bids are requested by invitation to a selected group of contractors. The invitation conveys much of the information that would be included in an advertised bid package.

Bidding Requirements

Bidding requirements for a company may be spelled out using a general provision section like the one used in a standard contract. The main points to be covered are the qualifications required of a contractor, the preparation of the proposal, the proposal guarantee, a noncollusion affidavit, and a statement on how the proposal is to be delivered:

- *Qualification*—For a bid to be acceptable, the bidder must either be prequalified by the buyer or must furnish financial or other information to prove the capabilities to perform the work. A license (required in some states) may be all that is necessary.

- *Preparation*—The proposal should be laid out on forms provided by the buyer to avoid irregularities on the part of bidders that could nullify the bid. These proposals should be signed, notarized, and sealed into envelopes provided by the buyer.

- *Proposal Guarantee*—A bond may be required (usually 10 percent of the estimated contract value) to ensure that the contractor can do what has been stated in the proposal. The bonds posted by the unsuccessful bidders are returned.

- *Noncollusion Affidavit*—A noncollusion affidavit is generally required by public agencies but should also be added to commercial bid requests. This document states that the vendor will not collude with another vendor, or with a site employee, to obtain the contract.

- *Delivery of Proposal*—A bid may be delivered by mail or messenger, but it must be received before the time set for opening. Otherwise, it may not be accepted.

Most bids are evaluated based on price alone. If other factors concerning the job are important (such as technical competence), the bids may be evaluated with a predetermined point system. Once a decision is made, the buyer officially notifies the successful bidder that the bidder has been awarded the contract. The bidder is then expected to execute the contract agreement within the specified time.

Standard Forms of Construction Contracts

Contracts used in the United States have a standard form. Most contracts include the divisions shown in Figure 2-10.

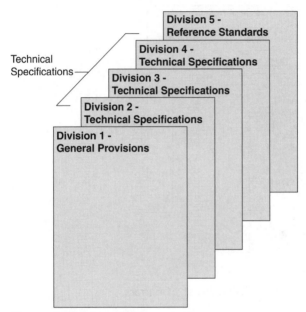

Figure 2-10   Standard Contract Divisions.

The general provisions or general conditions are included in the first section of the contract. A number of technical specification divisions may follow for each major part of a project. Applicable standards should be cited in the last section of the contract.

General Provisions of a Construction Contract

The general rights and responsibilities of the parties to a contract are spelled out in the general provisions. This language in a contract is sometimes referred to as boiler plate, or the legals, even though no part of the contract is required by law. The general provisions of a contract are shown in Figure 2-11.

The general provisions in a contract may not have all of the sections shown in Figure 2-11, and some contracts may use different

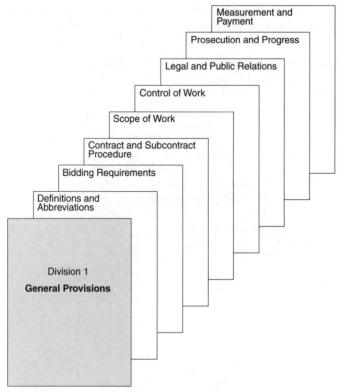

Figure 2-11 General Provisions of a Contract.

terminology, but the basic idea of this section in any contract still holds.

Definitions and Abbreviations
This includes definitions of terms and abbreviations used in the contract.

Bidding Requirements
These are instructions included in a bidding document, such as the invitation or advertisement for bid, instruction to bidders, bid form, and bid bond.

Contract and Subcontract Procedure
This section (found in both bid documents and contracts) should include a discussion of the contract bond and progress schedule

requirements. A description of how contracts will be awarded and how the contract will be executed can be included in this section. A subcontracting provision (or provision on whether or not a contract can be assigned) will usually be part of this section. A description of recourse for failure to properly execute the contract will also be included.

Scope of Work
The scope of work is made up of statements that describe the work to be performed, cleanup requirements, and final acceptance. Availability of space and storage for the contractor should also be discussed in this section. Permissible deviation from estimated quantities of labor and material should be spelled out with a statement in this section.

Control of the Work
This section will include the names of site personnel responsible for contract completion, a procedure for handling change order requests, inspection procedures, and acceptance requirements (if not already spelled out in the scope of work). Field office and other facility requirements may be included. Rules for relations with other contractors should be included if more than one competing contractor will be on site. Control requirements for site documents and plans should be spelled out. Receipt, inspection, tagging, and handling construction materials at the site should also be addressed in this section.

Legal and Public Relations
The general provisions of a contract will also include legal and public relations provisions to identify the legal requirements between the contractor, buyer, and general public. The purpose of this section is to ensure that liabilities for actions arising from the work are assigned to the proper party and that this party should be held accountable to make good all claims.

- *Damage Claims*—Wording that indemnifies and holds harmless officers and employees of a company is common in a construction contract. *Indemnification* is an obligation, assumed or legally imposed, on a party to protect another against loss or damage from stated liabilities. The words *hold harmless* may also be used to state the contractor's obligation in this matter. Damage claim provisions apply to all associated subcontractors as well. Essentially, the work is performed entirely at the contractor's risk. The contractor is usually required to provide insurance to cover this provision, since the buyer cannot count on the financial capability of the contractor. This insurance will include the following:

- *Worker's Compensation*—In statutory limits.

- *Comprehensive General Liability (Property Damage, Contractual Liability, and Personal Injury)*—$1,000,000 is common.

- *Comprehensive Automobile Liability*—$1,000,000 per occurrence is common.

Often the contractor is required to name the buying company as an additional insured to a policy.

- *Laws, Ordinances, and Regulations*—The contract must follow all applicable federal and state laws as well as local ordinances. Any licenses and permits required to perform and complete the project must be acquired by the contractor.

- *Responsibility for Work*—The contractor is responsible for material and equipment used while performing on the contract. No claims are allowed against the buyer for damages to this material or equipment. The contractor must make good all damage caused by the contractor.

- *Sanitary Provisions*—The contractor may be required to maintain sanitary facilities for personnel.

- *Public Safety and Convenience*—The contractor should conduct work so as to inconvenience the public as little as possible. Steps should be taken to install temporary crossings, take measures to prevent deposits of earth or other materials on roads, and to keeping flying dust to a minimum.

- *Accident Prevention*—The contractor must observe all the safety provisions and rules required of current site personnel. The contractor is responsible to provide safe working conditions on the project.

- *Property Damage*—The contractor is obligated, when using private property, to correct any damage they make.

- *Public Utilities*—The contractor must be advised to take care around public utilities such as water, natural gas, and electricity. All costs associated with avoiding and temporarily moving these utilities are the responsibility of the contractor.

- *Abatement of Soil Erosion, Water Pollution, and Air Pollution*—Contractors must minimize soil erosion and muddying of streams, irrigation systems, impoundments, and adjacent lands. Fuels, lubricants, and other liquid materials must not be discharged into adjacent waters.

Prosecution and Progress

Another part of the general provisions should deal with how work will start and proceed, the point of completion, suspension of the work, unavoidable delays, default of contract, liquidated damages, and extension of time:

- *Commencement and Prosecution of the Work*—The date the work is to start and the date it must be completed should be spelled out. Construction should progress in a manner that ensures completion according to a progress schedule. A statement should be made as to whether operations will be limited at the site including traffic and work by others.

    A statement should also be made about the required ability, adequacy, and character of workers the contractor employs and about the sound basis of the construction methods and equipment used by the contractor.

- *Time of Completion*—The number of calendar days from date of commencement should be spelled out rather than working days. The buyer may specify a date or time that they will begin using the constructed item, even if it is partially completed. A particular feature of the work that must be completed to allow subsequent jobs to progress should be spelled out here. This is especially important if the work described in this contract is only part of a larger project.

- *Suspension of the Work*—The buyer should be able to stop all work by the contractor if they feel it is necessary. So as not to breach the contract or prevent the contractor from fulfilling the contracted obligations, the buyer must identify all situations that may permit the buyer to stop work. Conditions such as weather, a strike, failure to perform up to the expectations of the buyer, or unsafe conditions or acts caused by the contract employees are all commonly identified.

- *Unavoidable Delays*—The contractor may be granted an extension in the contract time under certain conditions. However, the contractor is not necessarily entitled to compensation unless specifically spelled out in this section.

- *Annulment and Default of Contract*—The contract can be terminated under annulment or default. Annulment is when court order or plant management stops all work. The contractor is usually compensated for all costs incurred to stop work and make the site safe. Default occurs on the part of the contractor when the project is delayed or abandoned unnecessarily. Default can also occur if the contractor willfully violates terms of the contract or carries out the contract in bad faith. The buyer may use all contractor-furnished material and equipment to complete the project. Any bonds posted by the contractor may be used by the buyer to hire another contractor and complete the project.

- *Liquidated Damages*—The contractor is to pay the buyer for each day of delay in completing specified stages or the complete contract beyond the dates due. This may provide an incentive to the contractor to finish on time.

- *Extension of Time*—Just cause may exist for an extension of the stipulated time for completion of the project. This may include change orders, suspension of work, or delay of work for other than normal weather conditions.

Measurement and Payment

The general provisions of a contract should include measurement of amounts of the completed work and the payment of each level:

- *Measurement of Quantities*—All completed work of the contract will be measured for payment according to United States standard measures.

- *Scope of Payment*—Payment for a measured quantity for a unit-price bid will constitute full compensation for performing and completing the work and for furnishing all labor, materials, tools, and equipment.

- *Change of Plans*—This occurs when the measured quantities of work completed or materials furnished are greater than or less than the estimated quantities originally proposed.

- *Payment*—A procedure for partial payment should be laid out for longer jobs. The buyer may retain a percentage of the amount of the contract, pending completion of the contract.

- *Termination of Contractor Responsibility*—Upon completion and acceptance of all work included in the contract and payment of final certificate, the project is considered complete and the contractor is released from further obligation and requirements.

- *Guarantee Against Defective Work*—A guarantee period should be established for all or portions of the work, together with an amount of guarantee, usually calculated as a percentage of the contract cost. A guarantee bond may even be requested.

Technical Specifications

Following are three basic forms of technical specifications in a contract:

- *Materials and workmanship specifications*—Almost always found in a contract, materials and workmanship clauses cover the general and special conditions affecting the performance of the work, material requirements, construction details, measurement of quantities under the scheduled items of work, and basis of payment for these items.

- *Material procurement specifications*—When many separate general construction contracts are in force at the same time, a material procurement specification is usually added. This is especially true if some of the contracts are for similar types of work. This section would include a statement about the fabrication processes, as well as all the elements of materials and workmanship specifications (except for the field construction details).

- *Performance specifications* — Performance specifications are used mostly in procurement contracts for machinery and plant operating equipment.

Technical descriptions may be made for each functional area of a construction project, as shown in Figure 2-12.

Description

Under this heading, a concise statement is made of the nature and extent of the work included in the section and its pertinent features, including the general requirement that work conform to the plans and specifications.

Materials

When it is important to the quality of the work, materials of construction should be specified in a contract. The principal properties to be considered in the preparation of specifications of materials for construction are the following:

- Physical properties, such as strength, durability, hardness, and elasticity
- Chemical composition

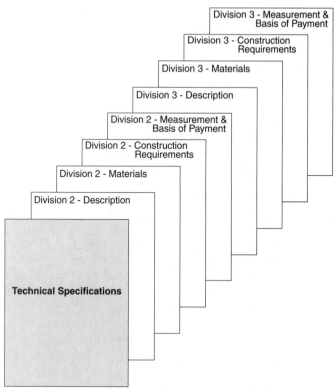

Figure 2-12   Technical Specifications of a Contract.

- Electrical, thermal, and acoustical properties
- Appearance, including color, texture, pattern, and finishes
    Inspections, tests, and analysis made by the manufacturer of the material should be acquired by the contractor and later provided to the buyer. Sometimes the Materials and Workmanship section of the contract covers all that's required for materials specifications. The words *or equal, or approved equal,* and *or equal as approved* are often used if a precise standard may be too restrictive.

Construction Requirements
This section is a detailed description of requirements sited in the General Provisions—Control of the Work section and of all the work required to complete the project.

Measurement and Payment

This is a more detailed accounting of the payment method described in the General Provisions—Measurement and Payment section as it applies to a specific part of the project.

Reference Standards

Reference all applicable standards for the job in this section. Watch out for standards that contradict each other. This mistake is commonly cited as a reason to ignore the standards listed. Common standards cited are the American Society for Testing and Materials (ASTM), American National Standards Institute (ANSI), American Society of Mechanical Engineers (ASME), and so on.

Summary

A shutdown planner should be sure to allow enough time to plan the shutdown. Additionally, the shutdown planner should make assignments to other individuals to spread out the responsibilities.

Tracking long-delivery items and delivery of large items is one key to shutdown management. Persistence of the buyer is not lost on the supplier. Common milestones to monitor are engineering, design assignment, design completion, fabrication, performance testing, shipment packaging, and delivery date. Make the supplier report the impact that a missed milestone has on future milestones.

Standard accounting practices still hold true during a shutdown. It's a good idea to assign a number to the shutdown, which should be marked on all purchase orders, work orders, stores slips, and contracts associated with the shutdown.

An organized thought process should be used to plan each job in the shutdown. The planning thought process has three main purposes: establish an estimate of time required for the job, define understandable steps to complete the job, and identify the materials, parts, tools, and equipment required for the job.

Break a job down into a number of steps and ask the same questions about each step:

- What must happen first on the job?
- Who must do this step?
- How many people are required?
- What parts, materials, or supplies will be needed?
- Is any support equipment required?
- How long will it take?
- What must happen next on this job?

A simple planning worksheet helps when recording each step for the job.

When applying the planning thought process to large jobs, a planner may classify the segments of a job. A large job will proceed through the following phases:

- Assignment and procurement
- Travel and staging
- Equipment clearance and entry
- Scope of work
- Return to service
- Clean-up and travel
- Closure and documentation

The activities required by operations should also be included in the jobs required for the shutdown. Estimates for the time required to shut down, decontaminate, or lock out equipment in the affected area should be solicited to improve the overall performance during execution.

A formal work package should be developed for activities that may be performed infrequently or for details for contractors. A collection of documents required to perform a job should include the work order, a spec sheet, tool/parts lists, sketches/photos, prints/drawings, a list of needed permits, and excerpts from equipment manuals.

A contract should be drafted whenever construction or other services are required. The form of the contract can be informal or may require a bidding process along with detailed descriptions of the work and expectations of how the work will progress.

Chapter 3 discusses how all of the planned tasks should be scheduled in a shutdown.

Chapter 3

Scheduling the Work

The scheduling phase of project management can vary from job to job and from facility to facility. Just as no one form of management is best for every facility, no one scheduling method fits every facility. Scheduling methods tend to vary depending on the type of operation.

The scheduling phase determines the order in which work will be performed. The efforts in this phase can utilize a variety of methodologies, several of which are discussed in this chapter, including the following:

- Critical Path Method (CPM)
- Load leveling
- Project duration versus project cost
- Program Evaluation Review Technique (PERT)
- The Monte Carlo Method
- Project management software programs

Critical Path Method (CPM)

Large jobs (such as projects, shutdowns, turnarounds, or outages) require a level of organization not normally needed in day-to-day scheduling. Many different, but interdependent, parts of the job must be coordinated. The jobs could extend over days, weeks, and sometimes months. Ensuring that all interrelated tasks are performed on schedule, over extended periods of time, with many different crews performing the work, can be a harrowing task.

During the early days of project management, large scheduling efforts centered on the use of Gantt charts. Gantt charts attempt to display the project's key activities on a time line. Figure 3-1 shows a typical Gantt chart.

A bar on the right represents each activity described on the left. The bar length indicates the start time and duration of the activity.

The major deficiency of a Gantt chart is the inability to identify project activities that are interdependent. In other words, it may not be obvious on a large Gantt chart that one activity cannot begin until one or more preceding activities are completed. Interdependencies that are not obvious make it very difficult to identify and react to potential delays on large projects. This is especially true when a key activity is delayed and/or not completed on time.

Critical Path Method (CPM) was developed to satisfy this deficiency. CPM is a scheduling technique that was originally

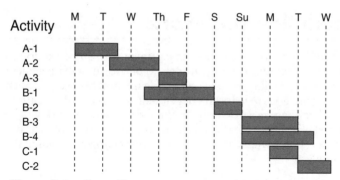

Figure 3-1 Gantt Chart.

developed in conjunction with the building of the Nautilus, the U.S. Navy's first nuclear powered submarine. The Nautilus project involved more than 200 contract firms. Coordinating the efforts of each phase of the project required that all the relationships between jobs performed by independent groups be known at all times. CPM logic diagrams clearly show the interdependency of all activities on other activities.

Logic Network Conventions
Critical path networks are represented through either the Arrow Diagram Method (ADM) or the Precedent Diagram Method (PDM). As shown in Figure 3-2, ADM (the traditional or first method for representing a logic network) identifies an activity as an arrow with circles, or events, noting its beginning and ending.

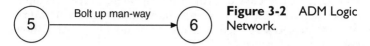

Figure 3-2 ADM Logic Network.

In the example, event 5 denotes an event, such as "Tank is now clean," and event 6 denotes the event "Man-way bolted shut." The beginning event is also the ending event for any preceding activities, and the ending event is the beginning event for subsequent activities. The line with the arrow represents the activity "Bolt up the man-way." The line and arrow usually point toward the right but the length of the line does not represent the passage of time or other resource.

Resources required to perform an activity are often identified on the activity line. Resources are definable inputs that must be allocated to complete activities. Common resources that can be controlled with a CPM logic network are labor, material, supplies, parts, equipment (cranes, lifts, and so on), or time, as shown in Figure 3-3.

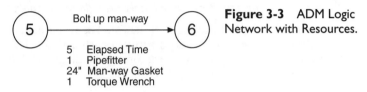

Figure 3-3 ADM Logic Network with Resources.

5 Elapsed Time
1 Pipefitter
24" Man-way Gasket
1 Torque Wrench

In ADM, the dependency of two activities is identified by the fact that both activities share an event. Suppose that, after the activity of bolting up the man-way is completed, a pressure test can be started. The fact that the pressure test cannot physically be started until the man-way is bolted shut would be diagrammed as shown in Figure 3-4.

Figure 3-4 ADM Logic Network—Interactivity.

The two activities share event 6, which is the ending event (5) of the activity of bolting up the man-way, and the beginning event (7) of conducting the pressure test.

The precedent diagramming method (PDM) dispenses with event identification and uses a box to represent the activity. Lines, drawn from left to right, represent the interdependencies of different activities. As with the ADM, these logic lines do not represent any elapsed time. The previous example of two dependent activities can be graphically presented in PDM as shown in Figure 3-5.

Figure 3-5 PDM Logic Network—Interactivity.

The means of determining the dependencies of all activities is the same in either diagramming method. Once all project activities have been identified and planned (all resources identified), the logic that ties them together is determined by establishing the precedent logic. This does not mean the activities have to be listed in order. Instead, the proper order is determined by asking one question for each activity in the list:

Which activity or activities immediately precede this activity?

This question is asked of each task in the list. The result defines the interdependencies and connections between each activity. This relationship is then drawn to reveal the precedent diagram. The following simple example describes this process.

Preparing a Logic Network

To prepare a logic network the discrete activities that make up the job must be identified. For example, consider the list of activities in Table 3-1.

Table 3-1   Example List of Activities

Reference	Task	Estimate
A	Frame and pour new pump foundation	8 hrs.
B	Erect Scaffolding	5 hrs.
C	Remove old pipe	6 hrs.
D	Mount new pump on foundation	8 hrs.
E	Install pipe hangers	3 hrs.
F	Install new pipe	8 hrs.

These activities are in no specific order. The optimum sequence for all these jobs depends on the resources available. If there are limited resources, and only one job can be performed at a time, the total duration of the project would be 38 hours. However, assuming that sufficient resources will be available to work some activities concurrently, the total duration of the project can be shortened.

The planner may review the list and develop a precedent logic as shown in Table 3-2 after asking the question, "Which activity or activities immediately precede this activity?"

With these logic statements we can develop a logic diagram. Figure 3-6 shows a sample logic diagram for these statements.

The preceding logic diagram represents all the logic statements. A computerized project program can also develop the logic network, but the user still has to identify precedent logic.

Finding the Critical Path

Project scheduling methods usually are used to determine the shortest amount of elapsed time required to complete a series of

Table 3-2   Precedent Logic

Reference	Precedent Logic
Activity "A"	Has no precedent.
Activity "B"	Has no precedent.
Activity "B"	Must be completed before activity "C" can begin.
Activity "A"	Must be completed before activity "D" can begin.
Activity "B"	Must be completed before activity "E" can begin.
Activity "C"	Must be completed before activity "F" can begin.
Activity "E"	Must be completed before activity "F" can begin.
Activity "D"	Must be completed before activity "F" can begin.

interdependent activities. There are often several paths of related activities that will start at the beginning event of the first activity and continue until the end event of the last activity. However, there is usually only one path requiring the most elapsed time from the start to the end.

For the earlier example, the critical path consists of activities A, D, and F. The total time required by the project when sufficient resources are available is only 24 hours. This is a great improvement over the time required to perform all the work sequentially (38 hours). If these critical activities are delayed in any way, the project time will be extended.

Other activities on the network can be delayed if necessary. On a simple network, such as the one in the example, jobs with delay can be discovered quickly. However, it is a good exercise to analyze the network for *float* or *slack time*.

The *earliest completion time of an activity* (T_E) is the sum of time for all the activities in a chain, up to and including the activity. The earliest completion times for the activities in our example can be determined as shown in Table 3-3.

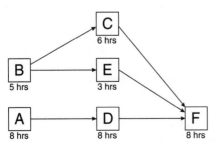

Figure 3-6   Sample Logic Diagram.

Table 3-3   Earliest Completion Time

Activity	Earliest Completion Time	T_E
A	A = 8	= 8
B	B = 5	= 5
C	T_E for B + C = 5 + 6	= 11
D	T_E for A + D = 8 + 8	= 16
E	T_E for B + E = 5 + 3	= 8
F	T_E for D + F = 16 + 8	= 24

Similarly, *the latest completion times* (T_L) are the latest possible times that each activity can be finished without increasing the length of the project. They are determined by starting with the total duration of the project and subtracting activity times until the first activity is reached. The latest completion time for each activity in our example should also be calculated as shown in Table 3-4.

Table 3-4   Latest Completion Time

Activity	Latest Completion Time	T_L
F	CP = 24	= 24
E	T_L for F − F = 24 − 8	= 16
D	T_L for F − F = 24 − 8	= 16
C	T_L for F − F = 24 − 8	= 16
B	T_L for C − C = 16 − 6	= 10
A	T_L for D − D = 16 − 8	= 8

Once this information is tabulated, float or slack time can be calculated. *Float* can be defined as the difference between the earliest completion time and latest completion time for each activity:

$$F = T_L - T_E$$

The logic diagram in Figure 3-7 shows the float calculated for the project in our example.

The earliest and latest completion times for the activities defining the critical path are equal. The resulting float for these activities is zero. The critical path, in technical terms, is defined as the series of connecting activities in a network that have no float.

Float can be used to develop a priority list for all activities not associated with the critical path. Activities with little float will be put near the top of the priority list, and activities with large float will be near the bottom of the list. Also, by moving the start time

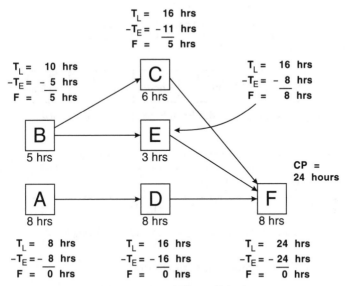

Figure 3-7 Logic Diagram with Float Calculation.

of an activity with float the planner can level a resource load. (Load leveling is discussed later in this chapter.)

Prioritizing work based on available float is also helpful in high-lighting off-site work. These activities can become out of sight, out of mind, especially if they have some float. Material that is on order or equipment that is out for repair may not be in the critical path. However, if the float time is minimal, any delay in the delivery can quickly make those activities part of a new critical path. If the float is exceeded, the completion time of the project will be extended and a new critical path will develop.

Interfering and Free Float
Total float for any activity represents the maximum delay for that activity before it becomes critical. *Free float* is the maximum delay of any given noncritical activity that does not effect the scheduling of any successor activities. *Interfering float* is the maximum delay of any given activity that affects the scheduling of any successor activities or diminishes the float of any predecessor activities. Figure 3-8 illustrates the difference between the two types of float.

The PDM network shows that the activities A, C, E, G, H, and I make up the critical path, whereas activities B, D, and F have some float. Figure 3-9 shows a Gantt chart version.

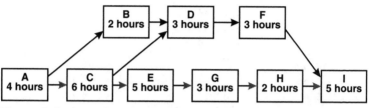

Figure 3-8 Logic Diagram with Both Free and Interfering Float.

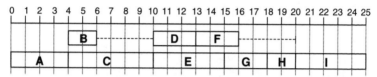

Figure 3-9 Gantt Chart Illustrating Free and Interfering Float.

In each of these versions, note the following:

- There is no float for critical activities.
- Activities D and F have 4 hours total float.
- Activity F has 4 hours free float. Any delay or extension under 4 hours will not affect the scheduling of activity I.
- Activity B has 8 hours of float. Four of those hours are free float. An additional 4 hours is interfering float that can affect the start of activity D.

Free float for any given activity can also be defined as the difference between the earliest start of a successor activity and the earliest completion of that activity.

Free Float = Earliest Start Time of Successor
– Earliest Completion Time

The earliest start time for activity D is 10 hours because activity C must be completed first. The earliest completion time for activity B is 6 hours, which is the duration of activity A plus the duration of activity B. Thus, the calculation for free float would be as follows:

Free Float for B = 10 hours – 6 hours = 4 hours

Activities with free float are sometimes referred to as *hammock activities*.

Load Leveling

A common dilemma often arises after a logic diagram is developed and the critical path is determined. Jobs that can be performed concurrently with other jobs require more resources than are available during certain periods in the schedule. Other jobs may be scheduled in a manner that inefficiently uses some resources (that is, a resource is scheduled to work every other day). *Load leveling* (also called *resource leveling*) is often required to solve problems of overallocation or inefficient use of resources.

Resources should initially be allocated to jobs in a project without regard for the obvious limit on the number of resources. In other words, assume you have unlimited resources. This obvious rejection of fact helps set the stage for a more realistic reallocation of resources based on the constraints of elapsed time and money. The following strategies are often used to resolve overallocated or poorly allocated resources.

- *Extend the project duration*—If the total elapsed time required to complete the critical path is less than the time allocated for the project, put off the start of some critical jobs.

- *Delay the start of noncritical jobs within their float*—This is usually the solution of choice if a resource is scheduled unevenly over the project duration.

- *Split noncritical jobs*—Schedule some jobs to stop and then restart when resources are available.

- *Expand the workday for the overallocated resource*—Increase the workday of some resources to a 10-hour or 12-hour shift schedule.

- *Increase the resources that are overallocated*—Assuming the initial limit put on the resource is correct, this can be accomplished by bringing in contract labor.

- *Use a support resource*—Determine if some work can be performed by another resource, which may be underutilized. Operations or other personnel can be employed as helpers or gophers for a skilled craft. (An existing labor contract may preclude this solution.)

- *Drop noncritical jobs*—Some jobs that are not part of the logic network should be dropped from the project. These are usually jobs that would have been nice to do if time or resources allowed. Unfortunately, time or resources may not allow. The jobs with the lowest preestablished priority should be dropped first.

• *Add noncritical jobs*—If some periods in the schedule under-utilize an available resource, add some noncritical, or nice jobs. The extra jobs with the highest preestablished priority should be added first.

Although computer-based project scheduling programs avail-able today offer automatic leveling commands, the process of lev-eling resources requires judgment and decision-making that are usually beyond the scope of these programs. The process of load leveling usually involves common sense—something only humans have.

Manual Load Leveling

The process of manual load leveling is relatively simple when the project activities or tasks are few in number. As the project becomes more complex, a more rigorous approach to load leveling is nec-essary. The following methodology will often satisfy overallocation problems:

1. Initially schedule all project tasks at their earliest start date and time.

2. Beginning with the first day (or other period), schedule all tasks that are critical (no available float).

3. If resources are still available, schedule the noncritical tasks with the least amount of float.

4. Continue to schedule noncritical activities until available re-sources are allocated.

5. Proceed in the schedule until resources have been freed up by the completion of any critical task(s) initially scheduled.

6. Schedule the next critical task(s).

7. If resources are still available, consider scheduling the next noncritical tasks with the least amount of available float.

8. Continue to schedule critical tasks as the precedent critical tasks are completed. Only schedule noncritical tasks if re-sources are available. Schedule noncritical tasks in order of ascending amount of float.

9. If the scheduling process causes an overallocation of resources, consider adjustments to noncritical tasks to free up resources for critical jobs. Make adjustments to scheduled noncritical jobs that have the largest amount of available float first. Make additional adjustments to noncritical jobs in descending order of available float.

10. Continue the scheduling process until all tasks have been scheduled.

The decision tree shown in Figure 3-10 also describes the process. Consider the project plan shown in Figure 3-11. With all jobs scheduled by earliest start times, the resource requirements vary from days with no resource required to a maximum loading of 12.

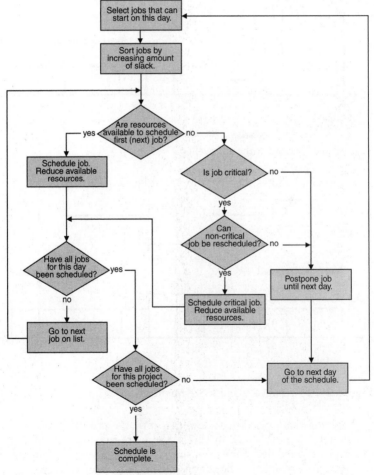

Figure 3-10 Float Decision Tree.

The schedule could not be worked as shown in Figure 3-11 if only six people are available.

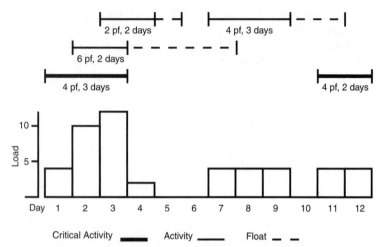

Critical Activity ——— Activity ——— Float – –

Figure 3-11 Load without Leveling.

Figure 3-12 shows the first attempt at leveling this craft.

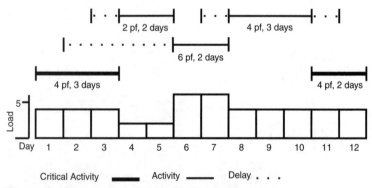

Critical Activity ——— Activity ——— Delay . . .

Figure 3-12 Load Leveled for Six Pipefitters.

Applying the logic of load leveling, only the critical job can be scheduled on the first day. By day 4, there are two noncritical jobs that can be scheduled. However, the available float on the one job requiring two people has been consumed. There are insufficient resources to schedule the other job until day 6. The fourth job is delayed in its start time until day 8. It will be completed by day 11, allowing the final critical job to be completed in time.

The pipefitter resource is utilized at or below its availability, but the scheduled use of the pipefitters is uneven. A planner may choose to leave this situation or to improve it through further leveling. Figure 3-13 shows the result of resource adjustments made on two of the jobs.

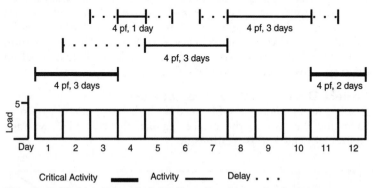

Figure 3-13 Load Leveled for Four Pipefitters (with Adjustment to Job Resources).

The job requiring two pipefitters was staffed with four to shorten its time. Staffing was reduced to four on the job originally requiring six pipefitters. This schedule not only provides a loading below the available work force but it also levels the requirement of pipe fitters to four throughout the project.

It is important to note that although the pipefitter craft is now balanced, some activities are delayed until the last moment. These activities no longer have float. These activities may be incorporated into the path including other jobs with no float. This, by definition, is the new critical path. The project duration may be extended if any of these jobs are delayed. The planner should watch the execution of these jobs closely and add more pipefitters if they start to fall behind.

Leveling Other Resources
Other resources that may be constrained are special services, special tools, and support equipment. Be sure the activities involving these resources are added to the logic network. Load leveling (or perhaps project delay) may be required.

Special service problems can arise in many forms. A common situation involves safety tests (such as gas detection in confined space).

These types of tests are usually performed in the early moments of the shutdown. If only one person is allowed to use the gas-detecting instrument, or only one instrument is available, the start of concurrent jobs may be delayed.

Shutdown work that requires large crane usage can create a particular difficulty. At the beginning of a large shutdown, many lifts may be required early in the project. The availability of these cranes should be tracked as a separate resource to anticipate any conflicts.

Knowing where to place a crane is one additional problem that often occurs. A crane making a lift may block the movement of another crane or keep other vehicles from moving from one location to another. The location of every crane should be discussed before the start of each shift to avoid conflicts.

Project Duration versus Project Cost

A shutdown planner may be faced with the economic decision of balancing the cost of the project against the cost of downtime. Figure 3-14 shows this relationship.

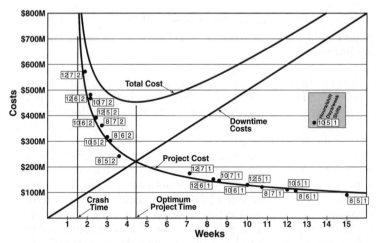

Figure 3-14 Cumulative Downtime (and Project Duration) versus Costs.

The cost of the project has been plotted against time, using different shutdown schedules. On the far left, a schedule of 8 hours per day, 5 days per week, and only 1 shift per day would take 15 weeks to complete but would cost less than $100,000. On the other end of the chart, a project schedule of 12 hours per day, 7 days per week, and with 2 shifts would take less than 2 weeks to complete but would cost about $570,000. Neither one of these scenarios may be the best possible schedule.

Also plotted on the chart are the cumulative downtime costs. The total cost of the shutdown is determined when downtime costs are added to the project costs. A few important points can be derived from this new curve:

- The lowest total cost of the shutdown is a little more than $450,000 and would take about $4^{1}/_{2}$ weeks (optimum project duration).

- Accepting a 10 percent increase in the lowest cost (just under $500,000) would shorten the shutdown to about 3 weeks.

- The most cost-effective schedule is not a 12-hour per day schedule, but rather a 10-hour two-shift schedule or an 8-hour two-shift schedule.

- The shortest duration for the shutdown, known as the crash time, is about $1^{1}/_{2}$ weeks.

Determining the actual crash time takes a little more effort than just extrapolating the project cost curve and involves maximizing the staffing and other resources for each job in the shutdown. This is still a worthwhile exercise because it can provide important management data. Very often, project managers are asked to complete the shutdown work more quickly, or even terminate all work and restart the plant. Outside economic factors are often cited as the reason for the change in plan. To the extent this is possible (some jobs cannot simply be stopped), management needs to know the cost of such a decision. The crash time estimates can be applied to the remaining jobs and a new cost can be quickly calculated using project management software. It is possible that upper management will rethink the decision once it sees the higher costs.

Modern project-management software programs allow the user to predict many possible outcomes for shutdown costs and duration. Key in this forecasting ability is the quality of the information entered into these programs. At best, the cost and duration data for tasks is based on past experience. However, much of the information must be based on estimates with little supporting data. Operations and other stakeholders would be severely misled if a planner entered only worst-case or best-case values. Some project management programs deal with uncertainty of estimates through probability theory.

Program Evaluation and Review Technique (PERT)

Some planners add additional hours to a job estimate to account for the possibility that an unanticipated problem will occur. In construction estimates, this is referred to as a *contingency amount*. The actual shutdown duration will usually be less than the estimated amount because many of the contingent events never occur. It can be argued that it is just as bad to finish a shutdown ahead of time as it is to finish late. Operations may have put off potential customers or furloughed workers, thinking the shutdown would last longer. Informing them you have finished early means they must mobilize quicker than they anticipated.

Program Evaluation and Review Technique (PERT) was one of the elements used to develop the CPM. The network developed by project management programs is often referred to as the PERT chart. Another (and often forgotten) part of PERT is the ability to forecast project completion times when the durations of individual activities are uncertain. Activity durations were fixed in the CPM examples used earlier in this chapter. The duration estimates are based on the judgment of a planner. Actual activity durations may be quite different, dependent on problems encountered during execution.

When uncertainty exists as to the duration of a given activity, the duration should be presented as a range. One method commonly used requires the determination of the following three estimates. (Other probability distribution methods will be described later.)

- T_o—Optimistic completion time
- T_m—Most likely completion time
- T_p—Pessimistic completion time

As the guidelines infer, T_o represents the activity duration time if everything goes right. It also implies that the problems commonly encountered will not occur. An example would be the repair of a leaky faucet. The optimistic completion time might be as little as 15 minutes and would imply that the shutoff valve under the sink actually works, the screw holding the valve assembly together isn't cemented to the valve body with hardened mineral deposits, and the replacement washer is, in fact, the correct one.

T_m, the most likely completion time, might be more accurately estimated at one half hour for this example. This duration recognizes the likelihood that the shutoff valve won't shut off the water entirely and allows for time to shut off the water main.

T_p, the pessimistic completion time, allows for all possible things that could go wrong to actually occur. In the example, the shutoff valve doesn't hold, the screw holding the valve together is broken while trying to break through the minerals that encrust it, and the shutoff valve below the sink starts to leak at the packing when the valve is reopened. This necessitates shutting off the water main once again so that the valve stem can be repacked. In this scenario, the whole job may run an hour and a half.

These three estimates are used to develop a distribution known as beta. With a beta distribution, the expected time (T_e) for completion of a specific activity can be calculated as follows:

$$\text{Expected Completion Time} = T_e = \frac{T_o + 4 \times T_m + T_p}{6}$$

Note that the T_m is multiplied by 4, giving it a greater weight. The standard deviation (SD) of the expected time can be calculated as follows:

$$\text{Standard Deviation} = SD = \frac{T_p - T_o}{3}$$

The 3 in the standard deviation formula reflects the assumed extremes of plus or minus $1^{1}/_{2}$ standard deviations from the mean. Assume the following estimates were made by a planner.

$T_o = 3.5$ hours
$T_m = 5$ hours
$T_p = 8$ hours

The expected completion time and standard deviation of this expected time can be calculated as follows:

$$T_e = \frac{3.5 + 4 \times 5 + 8}{6} = 5.25 \text{ hours}$$

The standard deviation of the expected time can be calculated as follows:

$$SD = \frac{8 - 3.5}{3} = 1.5 \text{ hours}$$

The T_e and SD should be calculated for all jobs in a project. When considering many interdependent jobs, each with its own uncertainties, the probability of completion must take into account the net effect of all the uncertainties. This is figured by calculating the variance (or v) of each job. The variance is the square of the standard deviation, as shown here:

$$v = SD^2$$

The example in Figure 3-15 demonstrates how a PERT evaluation can be conducted.

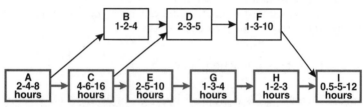

Figure 3-15 Project with Three Estimates.

In this network, the activities A, C, E, G, H, and I make up the critical path. There exists some uncertainty about these activities, which is reflected in the estimates shown in Table 3-5.

Table 3-5 Activities with Three Estimates for the Critical Path

Activity	T_o	T_m	T_p	T_e	SD	v
A	2	4	8	4.3	2	4.00
C	4	6	16	7.3	4	16.00
E	2	5	10	5.3	2.67	7.13
G	1	3	4	2.8	1	1.00
H	1	2	3	2.0	0.67	0.45
I	.5	5	12	5.4	3.83	14.67
		25		27.1	V =	43.25

The total expected time (T_e) to complete the critical path work is 27.1 hours. This value is slightly higher than the value that would have been calculated using the original estimates. However, we now know there is only a 50/50 chance the project will be completed in 27.1 hours. To find out how much we can improve our confidence in this prediction, we must first calculate the standard deviation of the total expected time.

The standard deviation of the total expected time is simply the square root of the variance total, as shown here:

$$\text{Total Standard Deviation} = \text{SD} = \text{square root of } V$$
$$= 6.6 \text{ hours}$$

The expected completion time and standard deviation of the total project are now known. The probabilities of completing the project within a specific period of time can now be calculated. This is determined by calculating a value called the deviation (or Z).

$$\text{Deviation} = Z = \frac{T_s - T_e}{\text{SD}}$$

where the following is true:

T_s = some completion date within the range

Once the deviation is known, Table 3-6 can be used to determine the probability of on-time completion.

Table 3-6 Probability Table for PERT Calculations

Negative Deviation	% Probability of Completion	Positive Deviation	% Probability of Completion
−3.0	0	0.1	54
−2.5	1	0.2	58
−2.0	3	0.3	62
−1.5	7	0.4	66
−1.4	8	0.5	69
−1.3	9	0.6	73
−1.2	11	0.7	76
−1.1	14	0.8	79
−1.0	16	0.9	82
−0.9	18	1.0	84
−0.8	21	1.1	86
−0.7	24	1.2	88
−0.6	27	1.3	90
−0.5	31	1.4	92
−0.4	35	1.5	93
−0.3	38	2.0	98
−0.2	42	2.5	99
−0.1	46	3.0	100
0	50		

Assume we would like to determine the probability of completing the project within 25 hours.

$$Z = \frac{T_s - T_e}{SD} = \frac{25 - 27.1}{6.6} = -0.3$$

From the table value corresponding to a deviation equal to 0.3, there is only a 38 percent chance of completing this project in 25 hours.

Now, assume that you would like to know the time that corresponds to a near certainty of completing the project on time (100 percent probability). From the table, this probability represents a positive deviation of 3.0, or 3 standard deviations. The following formula (transposed from the previous formula) can be used to find the time in question (T_s):

$$T_s = T_e + (Z \times SD)$$

For our example,

$$T_s = 27.1 + (3.0 \times 6.6) = 47 \text{ hours}$$

The result of this calculation means that the project can most likely be completed in 47 hours. This may be the time the planner wants to report to operations. If this estimate is unacceptable to operations (too long), the planner should review some of the jobs with the greatest variance and see if the pessimistic estimate could be improved through further defensive measures and contingency plans.

The previous example deals only with the critical path based on the most likely estimate. Other paths have the possibility of becoming critical should the worse case estimate for tasks in these paths actually come to fruition. Many modern project management programs employ a method that takes these possibilities into account.

The Monte Carlo Method

Project management programs that consider risk often provide the capability of using other distribution methods. In addition to the three-estimate method described earlier, normal, exponential, Poisson, binomial, uniform, and Weibull distributions can also be used. Some programs also offer the ability to enter actual durations from past activity. These entries are used to develop a distribution of future job duration.

Project software programs exist that expand on the basic statistical approach described earlier. These programs can provide a better distribution of possible project durations and costs and can show the relative criticality of each task. Questions such as the following can be answered using these programs:

- What are the chances of completing the project by a certain time or date?
- How confident are we that the cost for the shutdown will be below a certain level?
- What are the chances that a certain task will end up on a critical path?

Project programs capable of answering these questions require the user to provide either three estimates (as described earlier) or another type of distribution (such as a normal distribution, which is defined with only a mean and a standard deviation). Some can even accept recent job history and generate a distribution from that data. All this data is incorporated into an approach called the *Monte Carlo Method* to generate a truer distribution of total project duration and costs.

In the example used earlier, we assumed that only one path (A-C-E-G-H-I) would become critical under all possible combinations of job durations. This may not always be the case. Consider the data shown in Table 3-7.

Table 3-7 Three Estimates for All Activities

Activity	T_o	T_m	T_p
A	2	4	8
B	1	2	4
C	4	6	16
D	2	3	5
E	2	5	10
F	1	3	10
G	1	3	4
H	1	2	3
I	0.5	5	12

In this table, optimistic, most likely, and pessimistic estimates exist for all jobs. The pessimistic estimates on activities B, D, and F mean it is possible that two other paths (A-C-D-F-I or A-B-D-F-I) could become a critical path under certain conditions.

The method used earlier was also faulty in the fact that it assumed the distribution of possible project durations would be normal (that is, symmetrical). The actual distribution is more likely to be skewed to the left or right. Calculating the project duration for every possible job estimate would be very helpful in determining the actual probability distribution for the project. Unfortunately, this is not very practical. There are over one million possible combinations of these estimates for our simple example network.

The Monte Carlo Method (named for the French resort renowned for gambling and the invention of the roulette wheel) uses a unique sampling method to develop a representative distribution of project durations. Assume that a modified roulette wheel exists for each activity, as shown in Figure 3-16.

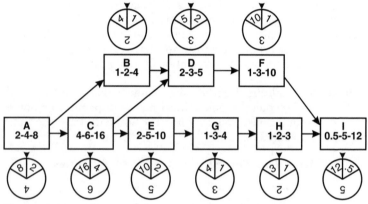

Figure 3-16 Network Diagram with Modified Roulette Wheels.

The computerized Monte Carlo Method obviously does not use a roulette wheel but rather a random number generator. The wheels in this example are marked with the optimistic, most likely, and pessimistic estimates for each activity. The most likely estimate has four times more area on the wheel to provide four times the chances. Next, all the wheels are spun to determine the first sample, as shown in Figure 3-17.

The result is one possible outcome of the project. The critical path has changed in this first try to A-C-D-F-I and the duration of the project is 24 hours (2 + 4 + 3 + 10 + 5).

If the wheels are spun multiple times, a distribution of the project possibilities begins to appear, as shown in Figure 3-18.

The final distribution (1000 samples) is a close approximation of the actual distribution of more than 1 million possibilities. We

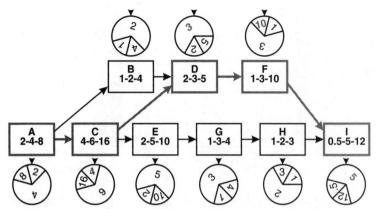

Figure 3-17   Network Diagram (Roulette Wheels Spun).

can now make some predictions based on the original estimates provided. We can say with confidence that there is only about a 7 percent chance to finish the project in less than 20 hours. Our chance of finishing improves to 40 percent (the sum of the first three probability bars) that we will finish in less than 25 hours. A chart can be derived from the data to represent the cumulative probabilities we are looking for, as shown in Figure 3-19.

From the previous chart you can see that the chance of being right improves to 90 percent if we say we can finish the job in 36 hours.

Another question a project manager might want answered is, "What is the probability that a task will become part of the critical path?" This information can also be calculated using project programs that include PERT calculations, as shown in Figure 3-20.

Activities A, C, E, G, H, and I have the highest probability of becoming critical with respect to project time. Activity B has almost no chance of being critical.

You may want to review any job in your project that has more than a 50 percent probability of becoming critical. These jobs will most likely dictate the duration of the project.

The true PERT method of project management may seem a bit cumbersome at first, but it may be worth the extra effort. Let's face it, if you're betting your job on a project schedule, you should at least know your odds of being right.

Computers have proven invaluable in shutdown and project management. Next we will discuss the modern project management

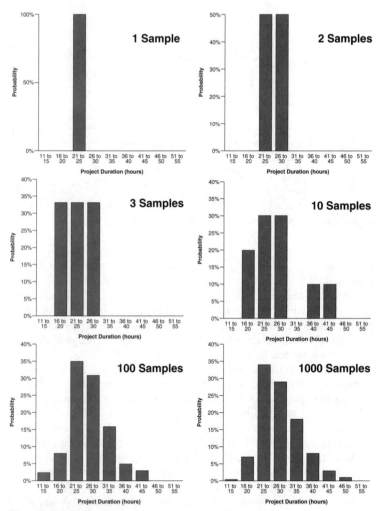

Figure 3-18 Successive Histograms.

software programs that make critical path and PERT calculations more accessible to every shutdown manager.

Project Management Software

Built around the idea of precedent logic, project management computer programs can make the task of developing a logic network much easier. Additionally, they can help the user to organize and sort

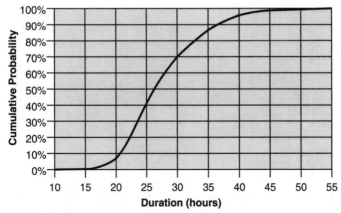

Figure 3-19 Cumulative Distribution.

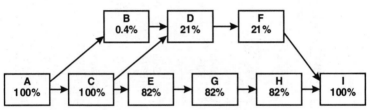

Figure 3-20 Probability an Activity Will Become Critical.

the many interdependent tasks. Resources with different working hours, schedules, and costs can be assigned to these tasks and the results can be viewed immediately.

Not only are these programs helpful in the planning phase, they are also a tool for analyzing and communicating project activity. The progress of the shutdown can be updated and tracked using these programs, and the costs can be monitored.

Choosing the Right Program

Complex and sophisticated programs are available and have been available for years. However, the simpler the program, the quicker someone can start receiving the benefit.

Table 3-8 shows a list of project software. This is not a complete list, but it is a good sample of programs that can be used in maintenance project scheduling.

Many of these programs cost less than $1000. Microsoft Windows is the most common platform for these programs, which

Table 3-8 Project Programs

Program	Company	Address	Phone	Web Site
Less-expensive project programs:				
Microsoft Project	Microsoft Corp.	One Microsoft Way, Redmond, WA 98052-6399	800-426-9400, 425-882-8080	http://www.microsoft.com/project
SureTrak Project Manager	Primavera Systems Inc.	Three Bala Plaza, Bala Cynwyd, PA 19004	610-667-8600, 800-423-0245	http://www.primavera.com
Higher-end project programs:				
Primavera Project Planner (P3) for Windows	Primavera Systems Inc.	Three Bala Plaza, Bala Cynwyd, PA 19004	610-667-8600, 800-423-0245	http://www.primavera.com
Open Plan Professional for Windows	Welcom Software Technology	15995 N Barkers Landing, Suite 350, Houston, TX 77079	281-558-0514, 800-349-8854	http://www.wst.com
Artemis Views	Artemis Corp.	4041 MacArthur Blvd., Suite 260, Newport Beach, CA 92660	949-660-7100. 800-477-6648	http://www.artemispm.com

makes them even more accessible. Low-end programs are sufficient if you are looking to get a fast start with computerized project management software. Intelligent on-line support is available in some programs. This feature can lead a novice through the process of project management without much difficulty.

Two other important criteria for selecting a project management program are the number of activities and resources used in the project. Small projects have less than 100 tasks and up to 30 resources. Medium projects contain 100 to 1000 tasks and up to 300 resources. Large projects are any task or resource requirement that exceeds the medium project limit.

For a small to medium project, Microsoft Project and SureTrak Project Manager are sufficient. Large projects may require Project Scheduler, Time Line, or one of the high-end programs. Most of the larger programs have built-in risk analysis functions (such as the PERT capabilities described earlier).

Sorting and Communicating Information

Most project management programs are designed for general business applications. However, many offer the ability to customize the program to your specific needs. Customized fields can be added for work order numbers or equipment numbers. These fields can be sorted, filtered, or searched quickly if the need arises. Many programs also employ a macro language that allows the user to automate operations (such as exchanging information with other programs).

Much of a shutdown coordinator's work is communicating with other people about a project. Project management programs can make this task easy, too. Following are a few of the features that are available:

- Logic diagrams (called PERT diagrams in many of the programs) and Gantt charts can be printed on demand. Although almost any printer can be used, this is best accomplished using a color printer or plotter as the output device. Tasks in the critical path can be displayed in a different color than the other tasks on a logic diagram. These charts can be annotated by the user to highlight certain sections.

- Reports can be customized to present project information the way the user wants it to be presented. Common canned reports available are baseline-to-actual, resource allocation, cost distribution, earned value analysis, and actual-to-budget.

- File import and export capability is built into most of the programs. Not only can reports be sent to a word processing file, data can be transferred to different formats. Many of the common formats, such as ASCII comma-separated text (.CSV), dBase (.DBF), Lotus 1-2-3 (.WK?), or Excel (.XLS), are available. Microsoft Project can be transferred to and from many programs through a format called Microsoft Project Exchange (.MPX).

- Database connectivity is available in some programs. One function, called open database connectivity (ODBC), lets external database programs share project information. Other programs allow sequential query language (SQL) database capability to be employed. The SQL option can allow the project to be updated in the project software and in a computerized database management system simultaneously. These functions will most likely have to be implemented by an information systems person at the site.

- Object locate and embed (OLE) client server functions allow you to drag-and-drop information from a spreadsheet program or word processor to the project program. OLE also allows project data to be linked to these programs and updated as it changes.

Most of the programs work on local-area networks (LANs) but only allow one user to update the project at a time. However, project information can be sent and received via email to inform project members and update the project database. This process is called *distributed* or *workgroup project management*.

Breakdown Structures

Project programs allow resources or activities to be organized in a number of ways. Many programs can display a work breakdown structure (WBS). The WBS is a set of codes used to identify activities in a hierarchical way. Many programs associate these codes with an outline structure as shown in Table 3-9.

The rebuild of the boiler has two different logical subsections. These subsections are further broken down into the specific work that will take place.

Another project program tool is organizational breakdown structure (OBS). This set of codes is used to identify tasks by resource groups. This structure is often used to reflect departmental structure in a company or code of accounts.

Not only can multiple resources and levels of resource be identified, but also each resource can have its own individual calendar. These calendars indicate the working days and hours for each

Table 3-9 Sample Work Breakdown Structure

3.0	Boiler rebuild	
3.1	Coal grate repairs	
	3.1.1	Remove grate hood
	3.1.2	Inspect grates
	3.1.3	Replace bad grates
	3.1.4	Inspect drive gear box
3.2	Tube repair	
	3.2.1	Remove boiler shell plates
	3.2.2	Inspect tubes
	3.2.3	Repair or plug bad tubes

resource or resource group. The whole completion of a project can be changed just by differences in available resources. Some programs have difficulty dealing with shift schedules for resources. If this is a major requirement, consider some of the higher-end programs.

Multiple projects can be worked simultaneously using project software. Projects often use the same people or other resources with overlapping assignments. Handling more than one project with a single project software file helps to sort out these situations. Some programs feature the ability to draw resources from a common pool for all projects.

Summary

Gantt charts were originally the tools of choice when it came to scheduling jobs in a project or shutdown. The start and duration of a job was indicated by the length and position of a bar on a timeline. Unfortunately, the interdependencies of jobs could not be easily seen. Critical Path Method (CPM) was developed to satisfy this deficiency.

In the CPM, a logic network of the discrete activities that make up the project is developed using precedent logic. The critical path (that is, the longest duration path) can then be derived. The jobs that make up the critical path also have no float (or slack).

Float can be used to develop a priority list for all activities not associated with the critical path. Activities with little float will be put near the top of the priority list, and activities with large float will be near the bottom of the list.

Jobs that can be performed concurrently with other jobs may require more resources than are available during certain periods in the schedule. Load leveling (also called resource leveling) is often required to solve problems of overallocation or inefficient use of resources. The following strategies are often used to resolve over-allocated or poorly allocated resources.

- Extend the project duration
- Delay the start of noncritical jobs within their float
- Split noncritical jobs
- Expand the workday for the overallocated resource
- Increase the resources that are overallocated
- Use a support resource
- Drop noncritical jobs
- Add noncritical jobs

The process of leveling resources requires judgment and decision-making that are usually beyond the scope of computer programs. A human touch is required.

Other resources that may be constrained are special services, special tools, and support equipment. Be sure the activities involving these resources are added to the logic network. Load leveling (or perhaps project delay) may be required.

A shutdown planner may be faced with the economic decision of balancing the cost of the project against the cost of downtime. The cost of a shutdown should be plotted against time, using different shutdown schedules. Additionally, the cost of downtime should also be plotted. The resultant chart will provide a guide as to what is the best shutdown schedule and duration. Additionally, the crash time (the shortest time in which a project can be accomplished) can be developed from this plot.

Adding hours to an estimate is not the best approach to cover contingencies in a job to account for the possibility that an unanticipated problem will occur. A better approach is to use Program Evaluation and Review Technique (PERT). Multiple estimates or a distribution of estimates are employed in PERT to answer key questions of uncertainty in a shutdown. Some questions answered are as follows:

- What are the chances of completing the project by a certain time or date?
- How confident are we that the cost for the shutdown will be below a certain level?
- What are the chances that a certain task will end up on a critical path?

Low-end programs are sufficient if you are looking to get a fast start with computerized project management software. The number of activities and resources that can be used in the project is another guide for purchase. Most of the larger programs have built-in risk analysis functions (such as the PERT capabilities described earlier).

Chapter 4 discusses the issues that arise during the actual execution of the shutdown.

Chapter 4

Executing the Project

The execution phase of a project or shutdown is, of course, the most important. If the plan is laid out well, all the allocated resources will be pulled together to realize their full potential. However, the shutdown coordinator should not just sit back and watch it all happen. Close monitoring of the work is just as important as the planning phase.

Nothing is more essential in leadership than knowing how and where to get information and then developing it, analyzing it, and knowing how to use it. Karl Von Clausewitz (the eighteenth century Prussian military strategist) pointed out that you can't ask a private what's happening on the battlefield because his fear, anxiety, or exhilaration consumes his attention and colors his information.

All that people report to the shutdown coordinator during a shutdown is data, not necessarily information. Everything is not equally important. It's up to the shutdown coordinator to cull the information out of the data. The following rules of management should be understood:

- Understand the vital few and the trivial many when receiving project data from others.
- Don't play hunches. Get the facts, then act accordingly.
- Fight personal prejudices. Just because you are normally an electrical supervisor doesn't mean the electrical work performed during the shutdown is the most important.
- Delegate or defer, but monitor the actions of people in your charge. You have the right to ask any question about the shutdown.
- Always return to the plan. If the critical path has not changed, you're on the right track.

This chapter examines some key topics related to executing the project, including the following:

- Shutdown safety
- OSHA requirements
- Developing daily schedules
- Updating job status
- Reporting project status
- Tracking project costs

Shutdown Safety

Construction sites are notorious for being unsafe environments for workers. Since the frenzy of activity that occurs during a turnaround, shutdown, or outage is a lot like a construction job, extra precautions should be taken. Many people may be working in the same area at the same time during the shutdown, so the potential of one group's work interfering with another's is possible. It's a good idea to step through all jobs that will be going on concurrently. The following questions of safety during this activity should be asked:

- Have all personnel (including contractors) been briefed on personal protective equipment requirements?
- Are safety procedures (such as lock-out, vessel entry, and hot work) understood by all?
- Are safety clearances needed for this job?
- Will adequate lighting be available?
- What type of lock-out will be used?
- Will blinds be required?
- Will people be working above other people?
- Will two or more groups be working in close proximity?
- Will a crane or other overhead device be used in this area?

A preshutdown walk-through with everyone associated with the shutdown is advised with very large projects. Check lighting requirements, check that there is free and open access to exits, and review any emergency plans.

OSHA has special requirements for crane activity, which should be especially heeded during a shutdown. If cranes are scheduled, watch out that jobs aren't scheduled under the crane's lift path.

OSHA Requirements

The Occupational Safety and Health Administration (OSHA) has some regulations that should be heeded during a shutdown that may not be required during normal operation. A shutdown coordinator should review these requirements with the plant safety manager prior to the shutdown. Following are just a few of these requirements:

- Contract personnel should be given an initial briefing at the site prior to their participation in any work. The initial briefing should include instruction in the wearing of appropriate personal protective equipment, what hazards are involved, and what duties are to be performed. Contract personnel should be availed all safety and health precautions provided to the employer's own employees.

- The Process Safety Management of Highly Hazardous Chemicals (regulation 29CFR 1910.119) states that, "The employer shall develop and implement safe work practices to provide for the control of hazards during operations such as lockout/tagout; confined space entry; opening process equipment or piping; and control over entrance into a facility by maintenance, contractor, laboratory, or other support personnel. These safe work practices shall apply to employees and contractor employees."

- Cranes are sometimes required during large shutdowns. If a large crane is not commonly used at your facility, it may be a good idea to familiarize yourself with the federal laws governing cranes. Both OSHA 29CFR 1910.179 and 1910.180 spell out the requirements of crane operation, loading, testing, and inspection.

  It is particularly important to identify overhead electrical lines in the path of a large crane. Review the areas of the crane operation and develop plans to avoid these lines.

Other OSHA regulations that may come into play during an outage include the following:

- *Confined Space*—29CFR 1910.146
- *Lockout/tagout*—29CFR 1910.147
- *Electrical Safe Work Practices*—29CFR 1910 Subpart S

Developing Daily Schedules

Some shutdown coordinators wrongly wait until the start of each shift to develop a schedule. They feel they won't know enough about the progress of jobs to develop a good schedule. Still others require the maintenance supervisors to develop their own schedule from the updated master schedule.

It is important that individual daily schedules be broken out of the master schedule before each day of the shutdown. It is best if this task is performed by or under the direction of the shutdown coordinator. Figure 4-1 shows a Gantt chart of a section of a master schedule.

The resources have already been properly leveled and extra downtime jobs have been added to better utilize the eight maintenance employees, eight per shift. Two 12-hour shifts are employed.

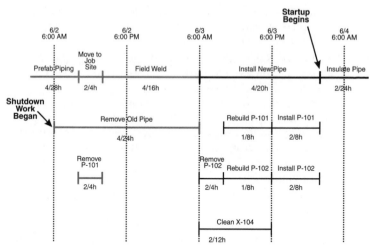

Figure 4-1   Master Schedule Gantt Chart.

The plant was shut down and ready for maintenance on June 2 and the old pipe was removed over the next 24-hour period. Concurrently, prefab work was completed, the pipe sections were moved to the job site, and the sections were welded into spool pieces of proper length in the field. A pump (P-101) was removed while the piping was being moved to the job site because two maintenance workers were available.

Assuming the field welding and pipe removal will be complete by morning, a schedule must be developed for the June 3, 6:00 a.m. shift. A tentative schedule should also be developed for the evening shift beginning at 6:00 p.m. The main job to schedule is the installation of the new pipe. Two extra jobs (Remove P-102 and Clean X-104) can also be started first thing in the morning because four other maintenance employees are available. P-102 will be removed by about 10:00 a.m. and the two employees on this job can be moved to rebuild both P-101 and P-102. Figure 4-2 shows a sample schedule.

The evening shift schedule can also be developed ahead of time. Any changes required to the evening schedule prior to the shift change should be minor. The maintenance first-shift supervisor can communicate these changes. First, the completion of the pipe installation is scheduled. Next the installation of the rebuilt pumps (P-101 and P-102) can be assigned to the remaining four maintenance

Daily Schedule

Supervisor	Day	Date	Shift
C. West	Monday	06/03/92	Day Shift - 6:00AM

W/O #	Equip. #	Description	T. Smith	M. Brewer	D. Huff	R. Smith	J. Stallcup	J. Bartlett	A. Meyers	A. Hanson	Days on Sched.	Orig. Est. Hours	Compl. Hours	Act. Hours Today
34526	R-100	Install new pipeline	12	12	12	12					0	80	0	
34545	P-102	Remove, rebuild, & install P-102				12	4				0	32	0	
35639	X-104	Clean X-104						12	12		1	24	0	
35572	P-101	Remove, rebuild & install P-101					8				0	32	8	
		Straight Time												
		Overtime												

Figure 4-2 Sample Daily Schedule.

workers. All six employees will be available to help with start-up problems eight hours later. If no problems exist, they can be sent home for the day. Insulalting the pipe can begin at that time or it can be rescheduled for the next morning when the plant returns to a regular shift schedule.

Updating Job Status

Up-to-date information is the key to project success. Lack of job information during project execution is the root of many project failures. The status of all jobs must be communicated in a timely manner.

A formal routine such as the following should be developed to report job progress:

1. Each morning, the shutdown coordinator should communicate with the shift coordinators. If a three-shift operation is being employed, the second (or night) shift coordinator should have already briefed the third (or graveyard) shift coordinator. Information provided by the coordinators should include the following:

   a. Delays and problems that occurred during the prior evening night.

   b. Which problems were resolved and which were not. Unresolved problems on back shifts to be handled by the day shift.

   c. Staffing changes that were made during the evening.

   d. The percent completion of all the jobs performed during the evening.

   e. A rough estimate of the elapsed time remaining on each job.

2. During the day shift, all problems that may result in a delay in completion time or an increase in project cost should be communicated.

3. At the end of every day shift, the shutdown planner should meet with all the project support personnel and update the project status.

4. Properly close out all work orders as they are completed. Don't wait until the end of the shutdown. This will help keep the shutdown cost accounting up to date.

5. Ask, "What heavy equipment coordination is required today?"

Reporting Project Status

A shutdown planner should be prepared to report the status of the project at any time. Early on in the project, it should have been determined who the customers or stakeholders are. The operations department is usually the most interested in the outcome of a shutdown because continued operations are at risk. The shutdown planner should determine what the operations information needs might be.

The status meetings held with shift coordinators and other project team members will provide the background information for all communications with operations. The shutdown coordinator should verify the statements made during these sessions and take actions to get the project on track. Once these actions are in place, the shutdown coordinator should conduct a session with the operations department.

Progress with respect to eventual start-up time is usually the most desired information. Any situations that may extend the project should be reported immediately. Any contingency plans and remedies should also be communicated to operations.

The project manager should also report if the shutdown work is progressing quicker than originally estimated. Operations may want to reschedule operators or other line personnel for the earlier date.

An atmosphere of total communication should exist during the shutdown. Other persons involved in the shutdown (such as maintenance workers) may want to know how their efforts are affecting the shutdown. The project master schedule should be displayed for all to see. Current status toward completion should be indicated on the schedule. Any benchmarks or goals that have been reached should be identified clearly. Figure 4-3 shows an example of a progress schedule.

A PERT chart provides the best overall look at project progress. The jobs that are completed are highlighted. Delays and extensions of certain jobs are also shown, along with a progress update. Despite some delays, this project is ahead of schedule. It can also be seen that jobs C and F became critical because of a 24-hour delay in the delivery of valves coupled with an early completion of jobs I and L.

Tracking Project Costs

Project cost data should also be published and displayed for all to see. As shown in Figure 4-4, there are two common ways to present the data: cost per week and cumulative costs.

Typically, the weekly charges associated with a shutdown or project start low, build to a peak, and then recede to zero. The curve

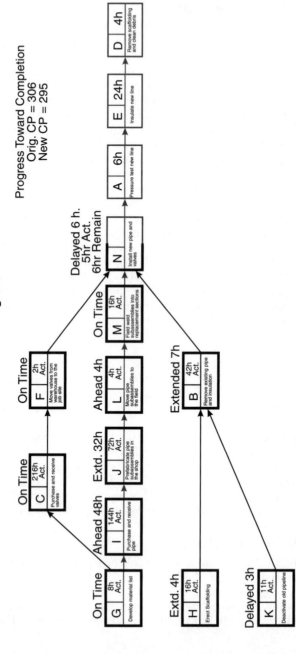

Figure 4-3 Progress on Shutdown.

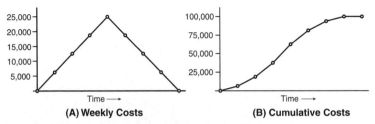

Figure 4-4   Weekly Costs and Cumulative Curve.

on the left shows an idealized example of this type of trend. The cumulative curve on the right shows how the same weekly costs accumulate to the total cost of the project. This curve is often referred to as an S-curve because of the shape it normally forms.

Budget versus actual costs is best tracked using the cumulative (S curve) and weekly charting method. Figure 4-5 shows an example of a set of actual and budget S-curves.

This chart shows progress made toward reaching the budgeted goal for the shutdown. Both estimated and actual costs are displayed. The initial phase of the project (which includes design, prefabrication, and material purchases) is shown for February through May. The scale changes to weeks for the project execution phase. Cumulative costs have exceeded the estimated costs so far. This cost overrun seems to be attributable to higher-than-expected preshutdown costs. Costs seem to be coming back into line but cumulative cost data, by itself, can be deceiving. Weekly costs and budgets should also be plotted to give a true picture of cost control. Figure 4-6 shows an example of such a chart.

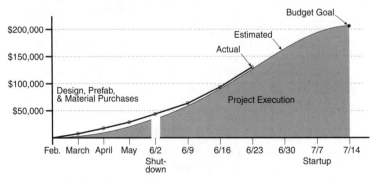

Figure 4-5   Cumulative Curve.

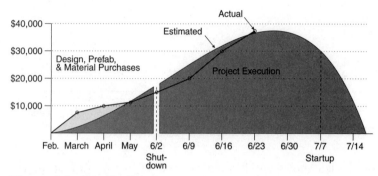

Figure 4-6   Weekly Costs.

This chart clearly shows the overruns in the preshutdown stage. It also shows that lower-than-expected costs were experienced in the first few weeks of the shutdown. What was not evident on the S-curve but is evident here is that the costs are again exceeding the estimated costs as of June 23. The shutdown coordinator should watch this trend closely and take steps to bring costs back in line if the budget goal is to be reached.

Summary

Nothing is more essential when managing shutdown execution than knowing how and where to get information and then developing it, analyzing it, and knowing how to use it.

The work site during shutdown should not be an unsafe environment for workers. Persons working in close proximity may interfere with other work groups. Contractors should be briefed on personal protective equipment requirements. Review lock-out, vessel entry, and hot work requirements prior to the start of work. Identify all safety clearances and lighting requirements in job descriptions.

The Occupational Safety and Health Administration (OSHA) has some regulations that should be heeded during a shutdown that may not be required during normal operation:

- *Process Safety Management*—29CFR 1910.119
- *Crane Operation*—29CFR 1910.179 and 1910.180
- *Confined Space*—29CFR 1910.146
- *Lockout/tagout*—29CFR 1910.147
- *Electrical Safe Work Practices*—29CFR 1910 Subpart S

It is important that individual daily schedules be broken out of the master schedule before each day of the shutdown. The schedule should be marked up and reviewed with the shutdown coordinator at the end of every shift.

Up-to-date information on all jobs must be communicated in a timely manner. A formal routine to communicate job status should be developed.

An atmosphere of total communication should exist during the shutdown. A shutdown planner should be prepared to report the status of the project to the stakeholders at any time. The status meetings held with shift coordinators and other project team members will provide the background information for all communications with operations. The shutdown coordinator should conduct a private session with the stakeholder.

Project cost data should also be published and displayed for all to see. The two common ways to present the data are cost per week and cumulative costs. Weekly charges associated with a shutdown or project start low, build to a peak, and then recede to zero. The cumulative curve shows the same weekly costs, accumulating to the total cost of the project.

Chapter 5 discusses the necessary reports and documents that must be developed to capture the essence of the shutdown effort. The documents developed will help plan, schedule, and execute the next shutdown.

Chapter 5

Reporting and Documenting the Shutdown Activity

The end of a shutdown or project is often accompanied by a sense of relief and disengagement. The project end should also mark a time for reflection and constructive criticism while the shutdown is still fresh in everyone's mind. Much can be learned by reviewing what went right and what went wrong during the shutdown.

A report that documents important results during the effort also needs to be prepared. This report must capture work that needs to be considered in the next shutdown. The preparation of the final report begins with a shutdown review meeting.

Project Review Meeting

Everyone that has assigned responsibilities during the shutdown should attend the project review meeting. Since this will include many individuals, it is often best to schedule the meeting to take place after regular work hours. Sometimes this review is best conducted off-site to minimize interruptions.

The goal of the project review meeting is to provide an open forum where general aspects of the shutdown can be discussed. In addition, it is at this review where specific assignments are made relative to written reports that will become the final report document and file.

Specifically, the project review meeting should cover the following topics that would be of general importance to the entire shutdown:

- *Safety/Environmental Performance and Procedures*
  - *Accidents*—All accidents should be discussed and recommendations made to ensure that none are repeated during future shutdowns.

  - *Near Misses*—Any near misses should also be discussed so that such occurrences don't become the accidents of the next shutdown.

  - *Permitting*—Any delays or problems associated with the acquiring of clearances on machinery or entry into lines or vessels should be discussed. Alternative plans to eliminate such delays or problems should be brainstormed. If such instances were numerous, a study group should be assigned to come up with proposals for future shutdowns.

- *Environmental Incidents*—Discuss any environmental considerations, such as emissions or hazardous waste handling.
- *Personnel Protection*—Identify any deficiencies in the availability or suitability of personal protective equipment. Develop recommendations for future shutdowns.

- *Contractor Performance*
  - *Training*—Identify any additional right-to-know requirements or means to minimize such training in future shutdowns by additional barricading or restricted work areas.
  - *Performance*—Identify any contractors who did not meet expectations or performance. Identify whether such nonconformance could be eliminated through tighter contracts, additional prejob site inspections, and so on.

- *Schedule Compliance*
  - *Schedule Breaks*—Identify all instances that caused schedule breaks, interruptions, or delays. Brainstorm innovative ways to minimize such disruptions in the future.
  - *Logistics*—Identify deficiencies in logistical areas. Propose other methods for delivery of materials, equipment, or supplies.
  - *Mobile Equipment*—Discuss problems associated with mobile equipment needs, scheduling, and placement. Give special attention to alternate lift paths or routing for heavy lifting equipment.

- *Work Assignments and Execution*
  - *Resource Allocation*—Discuss any unresolved problems with resource allocation. Specifically target jobs that were left incomplete or unfinished at the end of the shutdown.
  - *Execution Reports*—Assign report requirements to all who had specific shutdown assignments. A response date should be stipulated for these reports. Usually, two weeks is more than adequate. Allowing longer than this will invariably cause some individuals to put it off to the last minute. As time continues, the actual details of the shutdown will become clouded for such individuals.

- *Costs*
  Assign the responsibility of accounting for the shutdown costs. If a common account code or shutdown cost account was

established, collecting the costs is relatively easy. Take into account the costs that will continue to come weeks after the shutdown is completed. If a final accounting tally must be made, these remaining costs can be estimated. It is usually wise to estimate them higher rather than lower. If the actual costs come in lower than estimated, some department or cost center other than maintenance will be pushing to get the overrun corrected. If the actual costs come in higher than estimated, the maintenance department will have to exert its authority to have the corrections made.

- *Minutes*
  Prior to the project review meeting, assign someone the task of keeping accurate minutes of the meeting, including all discussions and recommendations. As an alternative, the meeting should be taped so that it can be transcribed later.

Project File

While the individual reports are being prepared, the project file can be started. This file should be the repository of all hard copy generated in connection with the shutdown. Specifically, the file should include the following:

- *All work orders initiated for the shutdown*—Set up subfiles for completed, unfinished, canceled, and new work
- *Contracts*—Include originals on all contracts for the shutdown
- *Insurance papers*—Include a file for all proof-of-insurance documents
- *Equipment reports*—Include original reports prepared for the work actually done during the shutdown
- *Drawings, prints, and sketches*—Include copies of all documents, or a listing of where these documents exist

Final Report

All of the information gathered and assigned at the project review meeting should ultimately end up in a final report. In addition to the specific details recorded at the meeting and the equipment repair reports subsequently prepared, the final report should include the broad description of the shutdown's major focus.

The final report should be a cogent document that is an accurate description of what was done, what needs to be done, and what did

Table 5-1 Total Labor Hours

	Estimated Hours	*Actual Hours*
Plant Labor	5103	4643
Contractor Labor	2730	2975

this all cost. The remainder of this chapter provides a sample report for a small shutdown.

Sample Shutdown Report

Following is a sample shutdown report for a distillation shutdown (SD# 032394-02) that occurred March 23, 1994.

Operations scheduled a 10-day shutdown of the distillation area for March 23, 1994. The maintenance department decided to take advantage of the downtime to perform needed repairs and modifications. Engineering also needed the time to tie in some environmental monitoring equipment.

Plant labor was scheduled for two 10-hour shifts as was contract labor. Ninety-two plant maintenance and 60 contract employees took part in the shutdown.

Table 5-1 and Table 5-2 describe the labor and material expended during this shutdown.

Planning for the shutdown was started three months earlier. It became evident that the two planners would not have enough time to plan all the work that could be performed during the five-day period. Other plant personnel were enlisted to help in job planning and material purchases. Table 5-3 shows a summary of major scheduled jobs, and Table 5-4 shows extra work to be performed if and when time allowed.

All major jobs were completed.

Table 5-2 Labor and Materials Costs

		Estimated	*Actual*
Plant	Prefab Labor	$182,143	$166,281
	S/D Labor	12,562	12,324
Contract	Prefab Labor	112,857	124,004
	S/D Labor	8,345	7,002
Materials	Storeroom	82,054	62,624
	Purchased	327,998	361,030
Rentals		7,020	6,341
	Total	$732,979	$739,606

Table 5-3   Summary of Major Scheduled Jobs

Process and Distillations Areas

1. Clean and repair X-450.
2. Clean/inspect CL-401, CL-402.
3. Replace CL-412 overhead piping.
4. Clean and repair leaks on CL-412 column.
5. Replace CL-401 overhead gas piping.
6. Replace damaged X-413 with X-407.
7. Install new design RA-452 and RA-454 relief piping.
8. Clean and repair holes on CL-432.
9. Repair leaks on CL-422.
10. Inspect packing in CL-441.
11. Repack CL-442, install new gas outlet.
12. Replace X-451, weak acid cooler.
13. Replace CL-427 tails cooler.
14. Clean X-404A and X-404C process sides.
15. Hydro-blast tubes on X-409.
16. Replace parts of CL-449 vent line.
17. Replace instrument air headers to CL-453 and CL-454 columns.
18. Replace CL-464 column, install new steel structure.
19. Replace X-450 backpressure motor valve (propane).
20. Repair numerous steam leaks on T-23.
21. Overhaul CL-401, CL-464 AP Inst., etc.
22. Pipe up VS-212A for weak acid.
23. Repack CL-200 column, replace bottom head, install extension.
24. Pipe up CL-200 column sulfuric acid cooler.

Utilities Area

1. Replace 6-inch main steam valve
2. Install piping for separating #3 and #5 boilers.
3. Install 8-inch orifice flanges in main steam line.
4. Repair deaerator internals.
5. Replace #4 boiler steam header valve.
6. Repack #5 boiler steam header valve.

Safety

One contract worker encountered minor acid burns. Otherwise no other injuries were suffered. No major safety violations were observed, although there was a problem with reporting the use of the

Table 5-4   Extra Work

Task	Complete?
1. Removal X-407, clean, install in Phase II (X-413).	Yes
2. Test and repair HA-331S.	Yes
3. Revise H2O piping on old X-413.	Yes
4. Tie in new CL-449 Column Tank.	Yes
5. Repair leaking nozzle CL-402.	Yes
6. Steam trace CL2 leads to Evap 1 and 2.	Yes
7. Repair CL-432 Product Line.	Yes
8. Install new CL-422 sight glass.	No
9. Clean CL-454, CL-412 reboilers.	No
10. Clean X-404A, B, C water sides.	No
11. Repair leak on CL-442 inlet.	Yes
12. Repair leak on CL-449 vent line.	Yes
13. Install new instrument air line to X-451 M/V station.	Yes
14. Repair P-406.	Yes
15. Pull, check, and reinstall P-408.	Yes
16. Replace valve on cooling water header to X-404's.	Yes
17. Reinstall 12-inch blind CL-402 top and remove.	No
18. Unplug suction lines to P-406 and P-408.	Yes
19. Replace angle valves Anhydrous Tank Car.	No
20. Hook-up gas driven pumps at pH basin.	No

fire water system for maintenance use. The safety supervisor wants to be informed whenever fire water is tapped. Some permit delays were encountered because only one gas sample unit is available in the plant.

Discussion of Major Shutdown Jobs
The following is a general discussion of the shutdown work (refer to Table 5-3 and Table 5-4):

 1. *X-450*—Nine pinhole leaks were found on the inlet tube sheet in the tube seal welds. All the tubes on the inlet tube sheet were rolled and a 50 psig N2 pressure test was performed. Two pinhole leaks were observed. These two tubes were rerolled and another pressure test was performed. No other leaks were detected. The exchanger was then recleaned, dried, and reassembled.

2. *CL-401, CL-402*—CL-402 top head was removed and the quench column trays were inspected to determine if cleaning was necessary. The trays were clean, the top head was reinstalled. The inlet to the CL-402 sparger was removed and the sparger was plugged up with carbon. All the carbon was chipped off and the inlet buttoned back up. The southeast 2-inch makeup nozzle on CL-402 developed a leak after the column had been reassembled. Thickness tests were done on both makeup nozzles and the column sides around the nozzles. The sides were okay, but both nozzles were very thin (.005 inch). Teflon sleeves with plastic steel on them were forced inside the 2-inch nozzles. Then, an Adams clamp was put around both nozzles. The leaking nozzle was also encased in water plug. The column was pressure tested to 20 psig and no leaks were detected.

CL-401 quench column inlet piping was removed and the column inlet sparger was cleaned. The 10-inch gas outlet from CL-401 was replaced. With the top piping removed, the column internals were inspected and it was determined that the trays were clean.

3. *CL-412 Column*—CL-412 column was cleaned and dried. Leaks on the top head were repaired. A temporary X-413 (X-407 from Phase I) was installed. The gas line from CL-412 to T-413 is 6-inch instead of 10-inch. No other major design conditions were changed. There is a pressure top and corrosion probe at the gas inlet and outlet and a temperature point on the gas/liquid outlet.

4. *RA-452, RA-454*—The new reactor relief piping was installed. All piping fit properly and worked sufficiently when heating the reactors and then starting the reactors. Two new rupture disks were installed and insulation was placed under both rupture disks to protect them from extreme abnormal temperatures that could cause premature failures.

5. *CL-432, CL-422*—The CL-432 column was set up, cleaned, and the leaks on the column were repaired. Considerable problems were encountered trying to get CL-432 dried so that it could be reassembled. Two nozzles were replaced. CL-422 column was air blown so the leaks on it could be welded up. No moisture was introduced here, thus drying was not required.

6. *CL-441, CL-442*—CL-441 top head was removed for inspection of the distributor and packing. The packing level in the

column was found to have dropped approximately 2 feet from where it was in October 1980. Approximately 3 feet of Teflon packing was added to bring the level back up to about 1 foot below the distributor. All other internals were in good condition.

CL-442 column was repacked with 31 feet of porcelain rings. The bottom head Teflon liner was damaged and was removed from the column. One piece of the porcelain packing support tray was broken and a piece of grating was cut and installed in its place (only for the end section). A new overhead gas line, 8 inch, was put in place of the badly corroded one.

7. *X-451*—The weak acid cooler was replaced with a rebuilt graphite exchanger. The old graphite domes were reused. A corrosion probe was installed on the water line outlet. New instrument air lines were run to the weak acid motor valve station next to the cooler.

8. *CL-427 Tails Cooler*—The tails cooler was replaced with a rebuilt graphite unit. A corrosion probe was installed in the outlet water line. New process line pipe hangers and supports were installed. The old tails cooler tubes were greatly reduced in size because of a silica buildup on the ID of the tubes.

9. *CL-464 Column*—The CL-464 column was removed from the structure so that a new support tower with working platforms could be constructed. The old B column was put back in place of CL-464 column because the older column had a weeping bottom nozzle. The two columns were identical. All piping on the Haveg vessel was modified to allow for an expansion joint at each nozzle on the column.

10. *CL-200 Column*—The column was repacked and a new bottom head with a Teflon liner was installed. The CL-200 extension was put on and was piped up for use. When the top head was removed, a large amount of packing was up in the overhead piping and in the distributor. Because of this, larger 3-inch porcelain packing was put on top of the porcelain saddles to help hold the smaller packing down during column upsets. All other column internals were in good condition.

11. *X-404A, B, C, and X-409*—These exchangers are of the falling film type (water on tube side). A hydro-blast unit was used to clean all of the tubes in these exchangers. It was found that great quantities of lime were built up on the tubes and a big improvement in cooling should be noticed.

12. *Cooling Tower Pump Check Valves*—All eight check valves were disassembled, inspected, and repaired where necessary. It was found that three out of four hot well checks were defective as were three of four cold well checks. All of the bad check valves were repaired by having a local machine shop rebush and repin the flappers. All flapper seats were in good condition. All eight check valves should now be as new.

13. *Deaerator*—The inlet baffle on the condensate return to the deaerator was welded back in place.

14. *Cooling Tower Fans, Gears, and Motors*—The oil in all three gearboxes was changed. Alignments of gear drive shafts were checked and corrected when necessary. New couplings were required on the West Cooling Tower fan and an oil seal on the West Cooling Tower gearbox needed to be repaired. An 18-inch butterfly valve was installed on the top inlet to the tower with new gear operator and hand wheel extension. Two cold well pumps were megohm checked below 1000 ohms and were sent out for repairs. Wetness was the cause of the insulation breakdown.

15. *Tower Internals*—The inside of the cooling tower was completely stripped. New design materials were reinstalled, which will help increase the tower efficiency. A great number of inlet dispersion nozzles were plugged with mud and lime. These were all cleaned and replaced. The tower was put in operation and the fan vibration level was checked. The vibration levels seemed above normal, but should by no means cause any immediate damage to the tower or the tower components.

16. *RA-452, RA-454 Heaters*—The heater systems for both reactors were repaired and at the same time modified to improve efficiency. By running a neutral buss and neutral wires to all sets of heaters, a failure on a main lead will cause three heaters to fail, where before all six heaters would be inoperable. Hopefully, this will double our efficiency and elongate overhaul intervals. Also, the electrical engineer had the opportunity to draw up a mechanical electrical drawing of both heater circuits. The engineer also did theoretical load calculations for the circuits and made changes that helped balance our loads. The actual load readings were within 10 percent of the theoretical.

17. *Main and Submain Circuit Breakers*—Four submain breakers and one 3000-amp main breaker were replaced during the

electrical outage. The breakers replaced were as follows: Phase I main, Chlorine building submain T, Phase II process area submain O, Cooling Tower sub main LV, and A submain for process area. The five breakers that were replaced will be sent out for complete rebuilds.

18. *Power Pole by Powerhouse*—A new pole was installed to carry the 13.8 kV power line conductors and static line. The pole was required because the old existing pole had a rotten top section, which could have broken and caused damage to transformers below the powerhouse and caused a plant power failure.

19. *Powerhouse M.C.C.*—The motor control center switchgear was inspected, cleaned, and repaired. The problem with the #1 boiler ID fan switchgear was corrected. The M.C.C. at the powerhouse is in a less-than-desirable location and an early spring overhaul is always required to ensure reliable operation through the hot summer months. Also, a new disconnect was installed for the emergency fire water pump. The damaged disconnect was found during the June outage of 1993.

20. *Cooling Tower M.C.C.*—Work done to the switchgear was as follows:

   a. All starters and breakers were removed from the enclosure, completely overhauled, and reassembled in the enclosure.

   b. The entire M.C.C. enclosure was sandblasted, repaired where possible, primed, and painted.

   c. New motor lead conduits were run and wire pulled to the Cooling Tower fan motors.

   d. All pump motors were megohm checked for insulation evaluation.

21. Plant and contract personnel were used to reinsulate RA-452 and RA-454. Approximately 150 feet of brine lines from P-408 and P-410 were insulated.

Comments and Recommendations

The shutdown went very well. All high-priority jobs were completed within the downtime allotted. The project planning software helped us discover some slack time in some crafts that allowed to us to do some additional shutdown and nonshutdown work.

Everyone got involved with the planning and actual work. Utilizing salaried and hourly people to plan and follow shutdown jobs really made a difference.

Following are some comments and recommendations received from various people:

- *The Safety Supervisor*—Any time we take both fire pumps out of service, we need a pumper from Lake Dreamland. I would like at least two days notice when we will need them so I can arrange for truck and notify the corporate office.

  Both sprinkler work and fire pump were not passed by me for approval as required for fire protection purposes. Fire loop and fire nozzles were used to flush cooling tower. One nozzle was damaged. Suggest that we provide separate equipment for this in future.

  Laborers were sent into tower with inadequate equipment (that is, boots). This resulted in an injury.

- *The Utilities Supervisors*—I think when the steam outage is planned, a more concentrated effort should be made to get jobs done in the Powerhouse area instead of not knowing if you are going to have enough manpower left over from the unit to complete the jobs that are scheduled. In this last shutdown, there was some question as to whether or not to cancel some of the jobs up to the last minute of steam outage.

- *A Maintenance Mechanic*—I suggest we find a more efficient manner of cleaning distillation columns and drying of same. For example, with CL-432 (that is, pipe steam to the column using $1\frac{1}{2}$-inch to 2-inch steel pipe), the hose effort takes too long to heat up the column. In drying, I suggest using a spare reboiler of such type that we could make portable and pass dry air through it and into the column as we did on CL-412, which I think provided positive results.

  Perhaps Shift Foremen and others could be better briefed on the scope of the jobs, decontamination, and procedures; possibly tentative scheduling of following certain jobs and responsibility of such projects as relating to preparation and so on.

- *A Maintenance Supervisor*—Remove reactor tops early in the shutdown and close to shift change.

  CL-412 column was washed instead of just air blown. Problems in drying caused delay in reassembly until final night of shutdown.

  Problems encountered with the propane system. Because of leaking valves, propane reentered the system, causing delay in blind installation. Torch should remain until maintenance is ready.

- *The Operations Manager*—Everything went very well during this shutdown. I'm sure the preplanning and help from different departments was the key. I was amazed at how much work was completed during this shutdown compared to ones we have had in the past.

Summary

The end of a shutdown or project should mark a time for reflection and constructive criticism while the shutdown is still fresh in everyone's mind. A report that documents important results during the effort also should be prepared. This report must capture work that needs to be considered in the next shutdown. The preparation of the final report begins with a shutdown review meeting.

The goal of the project review meeting is to provide an open forum where general aspects of the shutdown can be discussed, as well as make specific assignments relative to written reports that will become the final report document and file. Specifically, the project review meeting should cover the following topics that are of general importance to the entire shutdown:

- Safety/environmental performance and procedures
- Contractor performance
- Schedule compliance
- Work assignments and execution
- Costs
- Minutes

While the individual reports are being prepared, a project file (which should be the repository of all hard copy generated in connection with the shutdown) should be created and include the following:

- All work orders initiated for the shutdown
- Contracts
- Insurance papers
- Equipment reports
- Drawings, prints, and sketches

All of the information gathered and assigned at the project review meeting should ultimately end up in a final report, which includes the specific details recorded at the meeting and the equipment repair reports subsequently prepared. The Final Report should even include the broad description of the shutdown's major focus. This report should be a cogent document that is an accurate description of what was done, what needs to be done, and what did this all cost.

Appendix A

Boiler Shutdown Example

The following exercise documents the development of a boiler shutdown project. The boiler is a packaged unit (see Figure A-1).
The following statistics apply to the boiler and peripherals.

- *Design*—Water tube (350 tubes with two soot blowers)
- *Capacity*—80,000 lb/hr of steam
- *Working pressure*—300 psig
- *Design pressure*—500 psig
- *Dimensions*—30 ft long × 12 ft high × 10 ft wide
- *Fuel*—#2 fuel oil, 30,000 gallons aboveground tank

A packaged boiler is so named because the entire boiler is shipped to the site on a train or flatbed truck. The skids for this kind of shipment remain on the boiler even after installation. This boiler is commonly referred to as a D-type boiler because the water wall and convection tubes, along with the steam and mud drums, form the letter *D* when viewed from the front of the boiler.

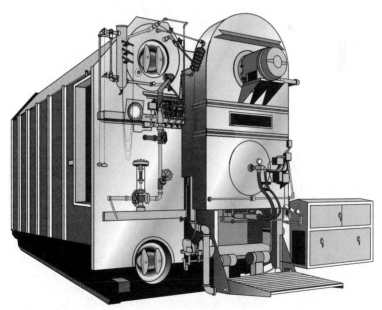

Figure A-1   Package Boiler.

The tubes are connected between the steam drum (top) and the mud drum (bottom). (Note that Figure A-1 is shown with a super heater, which may not exist in many boilers.) The large quantity of tubes directly below the steam drum is called the *generating bank*.

With the flame on, feed water enters the mud drum and is drawn by convection to the steam drum. Most of this upward convection occurs in the water walls and the back of the generating bank. Once in the steam drum, the heated water flashes to steam because of a reduction in pressure. The unflashed water returns to the mud drum through some of the cooler tubes in the left-hand wall and toward the front left of the boiler.

Any work performed on the boiler tubes (including plugging or replacing bad tubes) requires that a relatively small person enter the entryways in the steam drum or mud drum. Tubes are inserted into the steam drum walls and then flared into place. The tubes are mounted very close together. Any work to replace bad tubes often requires removing other good tubes.

A boiler requires extensive auxiliary equipment to support its operation. The process diagram in Figure A-2 shows the components in the system being discussed.

There are two fuel oil pumps (P-2) that supply #2 fuel oil to the burner. Only one is used at a time. Natural gas is required to maintain the pilot and compressed air (CA) is used to atomize the oil and improve the flame. A main blower (BL-1) provides combustion air. One of the boiler feed water pumps (P-1) is operated at all times. The other comes on only in low-water situations. The water is supplied through a deaerator system (DA-1 and T-1), which is used to remove air in the water. A set of condensate pumps (P-3) are used to pump returned condensate and deionized make-up water. Chemicals are pumped (P-4) from the mix tank into the mud drum.

Boiler History

The boiler was purchased and installed when the plant was first built in 1984. The unit was operated at less than half capacity for the first five years. As business began to build, the unit was required to meet increasing demand. Lax water-treatment practices that began in the early years led to serious problems in the last two years.

A planned shutdown two years ago included mechanical cleaning of all tube internals and intentionally plugging 23 tubes that were about to leak. This was acceptable for operations as the boiler continued to provide all the steam required by the plant. During this shutdown, emergency repair was required for the rear wall insulation, using plastic refractory. Some of the interlocking tile had been dislodged and was lying on the floor of the boiler firebox.

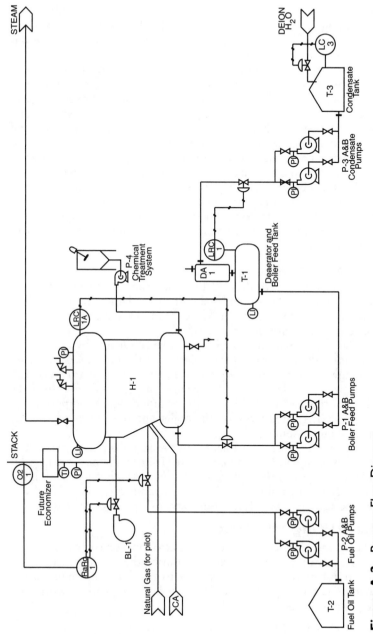

Figure A-2 Process Flow Diagram.

133

The furnace wall where the tile had fallen away was bulged out, indicating that the block insulation had pulled away from the clips. It was not immediately evident whether the refractory damage was a result of the burst tube.

Last year, a tube failure caused by pluggage forced a shutdown. An internal inspection revealed that 29 other tubes in the front of the generating bundle were about to fail because of excessive scaling and some bulging caused by overheating. The burst tube and all of the suspect tubes were plugged during the last shutdown, and the capacity of the boiler has been reduced by almost 20 percent. The result has been a lowered operating capacity in the process. There were two occasions when the operations department shut the boiler down because of low pressure and excessive water carryover. The lax water-treatment disciplines that caused the scaling problems have prompted plant management to pay more attention to the water treatment chemicals used and to the person in charge of boiler operation.

The Shutdown

Plant personnel have decided that they can no longer run at the low output of the plant and a shutdown is planned for July 7. The planner is told to take charge of the shutdown as a project manager, dedicating all work time to the effort.

The tube replacement will have to be conducted by removing a section of the rear wall of the boiler from the outside. Good tubes will have to be removed to get to some bad ones, so a total of 100 tubes will be replaced. A boiler repair contractor will most likely be used for this part of the shutdown because they are equipped to do all the code-related work on the boiler and have more experience than plant personnel.

The Engineering department wants to install an Economizer during the shutdown. This unit is mounted in line with the breech. Boiler feed water passes through tubes that have the hot gases from the boiler exhaust passing over them. This preheats the boiler feed water and saves energy, which would have otherwise gone up the stack.

Identifying the Work

The planner decided that a concerted effort must be made to identify all the work that can be performed during the shutdown. The plan was to use all the resources available to avoid missing an important item. The backlog file was also reviewed for pending work that requires a shutdown of the boiler:

- Change out P2-B
- Overhaul BFWP P1-A and B

- Check FD fan damper cams coming loose on the shaft or out of adjustment
- Change gasket on main header—steam leak
- Recondition BL-1 motor—low megohm reading
- Change zeolite in water softener
- Recondition P-3A condensate feed pump motor—low megohm reading

The shutdown report and file from the previous years was reviewed and the following items were listed:

- Initially inspect the firebox and tubing surfaces before cleaning the furnace. The boiler inspector would like to be present at this time.
- Wash the firebox and external tubing surfaces as well as the tube internals from the steam drum down to the mud drum drain internals. The boiler inspector would like to be present once it is clean.
- Allow at least 48 hours to cure refractory repairs.
- Don't let washdown sludge go into the process or storm sewers. This resulted in a fine last time. Have the sludge hauled away.
- Clean breaching to stack and inspect it prior to installing the new Economizer.
- Replace 100 suspect convection tubes. Turbine out the remaining 250 tubes.
- Check the boiler safety shutdown controls during the shutdown of the boiler and during the startup.
- Purchase new refractory to replace the patch on the rear wall that was put in two years ago.

The PM schedule was reviewed to take advantage of the total shutdown of the boiler. The overdue PM, as well as PM work that will come due during the months following the shutdown, were reviewed. The following PM work that requires shutdown of the boiler was identified for completion:

- Open and clean deaerator and thickness check tank
- Test and certify pressure relief valves
- Calibrate water level indicator and control loop
- Calibrate O_2 analyzer
- Overcurrent test motor O/Ls and C/Bs

- Megohm check boiler MCC starters, C/Bs, motors, and feeders
- Fuel oil tank inspection
- Overhaul flame sensor and control
- Clean and repack valve stems on 25 steam valves in the plant
- Replace thermal couples on the boiler, breaching, and stack

The following preliminary work was determined by the planner based on discussions with the maintenance supervisor:

- Sample and test fuel oil (bottom and top sample)
- Infrared scan boiler exterior
- Infrared scan MCC connections
- Vibration checks on rotating equipment
- Place purchase order with boiler repair and insulation contractor
- Check prints and purchase and receive boiler tubes
- Contact motor repair vendor
- Contact valve repair vendor
- Contact insurance inspector
- Purchase and receive fuel oil pump

The result of the infrared scan and the vibration checks revealed the following problems:

- Clean and balance forced draft blower BL-1
- Change breaker to P-3B condensate pump
- Tighten connection or change starter to BL-1

A walk-down of the steam system with maintenance management and some operations personnel resulted in the following jobs and comments about the pending shutdown:

- Rebuild the following control valves:
  - Deionized H20 control valve
  - BFW level
  - Fuel oil feed
  - Blower damper
  - Condensate feed
- Change out all pressure gauges on the boiler as well as the gauges on the FOP, BFWP, Cond P

- Drain and check chemical treatment systems
- Clean level indicator
- Clean level control electrodes
- Fix dent in stack caused by storm

A structured group interview was conducted with a mixed group of operations and maintenance personnel. The following tasks resulted from this effort:

- Calibrate ratio indicating controller on burner combustion system
- Calibrate level control on T-1 feed water tank
- Set aside more time for the hydro test of the boiler for repairs
- Infrared scan the steam and condensate piping as well as boiler external shell prior to the shutdown

Two nonshutdown jobs were also suggested to be performed when the boiler was shut down to take advantage of the fact that there was a contract insulating firm on site at the time:

- Repair insulation on condensate return line, main pipe rack
- Replace steam tracing on process feed line and reinsulate

These jobs can be performed if time permits.

Building the Work List

The original list of work was considered insufficient to lay out the entire shutdown, so the planner built a more detailed list. A work order was written for all activities performed by maintenance employees. A work order (as well as a purchase order) was also used for any material purchases or contract work. Accounting can easily collect costs charged against the work orders when the time comes to determine the cost of the shutdown.

Operation activities were identified to help keep the operations department better informed when their input is required. Permitting, lock-out/tag-out requirements, and shutdown and startup activities were noted. An estimate of the time required to perform these activities was also identified.

Some jobs were broken into segments to better utilize the load-leveling feature of the project management program to be used. For example, a simple pump replacement job can require a number of crafts. An electrician may be required to disconnect and reconnect the motor, operations may get involved in the lock-out, and the mechanic may get involved throughout the job. When all these people are assigned, a project management program assumes that all will be required for the duration. This is most likely not the case. Breaking the jobs into segments eliminates this problem in the program. Table A-1 shows the final work list.

Table A-1 Boiler Shutdown—Final Work List

Work Order #	Task	Duration	Resource
	Preliminary Work		
	Schedule I/R Scan Vendor	1 d	Planner
10117	I/R Scan of Boiler Ext.—PO# 5530	4 h	Mech, I/R Technology
10133	I/R Scan of Elect. Dist.—PO# 5530	4 h	I/E, I/R Technology
10125-1	Sample Fuel Oil Tank	1 h	Mech
10125-2	Analyze Oil Sample— PO# 5532	3 d	Oil R Us
10118	Vibration Analysis— Boiler Machinery	4 h	Mech
	Place P/O with Boiler Contractor	1 d	Planner
	Place P/O with Insulation Vendor	1 d	Planner
	Purchase and Receive Boiler Tubes	1 d	Planner
	Contact Motor Repair Vendor	1 d	Planner
	Contact Valve Repair Vendor	1 d	Planner
	Contact Insurance Inspector	1 d	Planner
	Purchase and Receive Fuel Oil Pump	1 d	Planner
	Boiler Shutdown		
	Shut Down Boiler	1 h	Ops
	Cool Down and Drain	23 h	Ops

(*continued*)

Table A-1   (*continued*)

Work Order #	Task	Duration	Resource
	Boiler Repairs		
10100-1	Open Firebox, Manways, and Breech	1 h	Mech[2]
10100-2	Confined Space Entry	1 h	Ops, Mech, Labr
	Firebox and Tubes		Insur, Planner
10100-3	Preliminary Inspection	2 h	Mech, Insur, Planner, Ops, Acme Boiler
10100-4	Water Wash Firebox and Tube External	8 h	Labr
10100-5	Water Wash Drums and Tube Internals	4 h	Labr
10100-6	Boiler Inspection	3 h	Ops, Insur, Mech, Labr, Acme Boiler
10105	Air Turbine 250 Tubes—PO# 5526	23 h	Acme Boiler
10106-1	Remove PRVs— Bring to Shop	1.5 h	Mech
10106-2	Test and Certify PRVs—PO# 5527	48 h	Valve Specialties
10106-3	Install PRVs	2 h	Mech
10107-1	Repl. 100 Conv. Tubes—PO# 5528	54 h	Acme Boiler
10100-7	Close Up Manways	1 h	Mech[2]
10111	Hydro Test Boiler @ 300 psig	8 h	Mech[2], Insur
10107-2	Repair Rear Wall Refr.—PO# 5529	20 h	Thermal Experts
10107-3	Close Up Firebox	1 h	Mech[2]
	Boiler Startup		
	Fire Boiler and Cure Refractory Repairs	48 h	Ops
	Process Startup	8 h	Ops
	Main Blower Repairs—BL-1		
	Lock-Out, Tag-Out of BL-1	0.5 h	Ops, Mech, I/E

(*continued*)

Table A-1 *(continued)*

Work Order #	Task	Duration	Resource
10104-1	Disconnect BL-1 Motor	0.5 h	I/E
10104-2	Remove BL-1 Motor—Bring to Shop	1 h	Mech[2]
10104-3	Recondition BL-1 Motor—PO# 5525	48 h	Short Electric
10104-4	Install BL-1 Motor	1 h	Mech[2]
10104-5	Connect BL-1 Motor	0.5 h	I/E
10104-6	Clean and Balance BL-1 Fan Wheel	1.5 h	Mech[2]
10122	Repair BL-1 Damper Cams	2 h	Mech
	Condensate Pump Motor Repairs		
10110-1	Lock-Out, Tag-Out of P-3A	0.5 h	Ops, Mech, I/E
10110-2	Disconnect P-3A Motor	0.5 h	I/E
10110-3	Remove P-3A Motor—Bring to Shop	1 h	Mech, C/Pkr
10110-4	Recondition P-3A Motor—PO# 5525	48 h	Short Electric
10110-5	Set P-3A on Base and Connect	1 h	Mech, C/Pkr
10110-6	Align P-3A Motor	2 h	Mech
	Economizer Installation		
5793-1	Inst. Scaffold at Breech—CAP# 123	2 h	Mech[3]
5793-2	Clean Breaching to Stack	1 h	Labr[2]
5793-3	Install Economizer in Breech	8 h	Mech[3], C/Pkr
5793-4	Install Economizer Piping	16 h	Mech[2]
5793-5	Pressure Test Economizer	4 h	Mech

(continued)

Table A-1   (continued)

Work Order #	Task	Duration	Resource
5793-6	Insul. Economizer Piping—PO# 5533	8 h	Thermal Experts
	Fuel Oil Repairs		
10126-1	Lock-Out, Tag-Out of P-2B	0.5 h	Ops, Mech
5254	Change P-2B Fuel Oil Pump— PO# 5535	2.5 h	Mech
	Empty Fuel Oil Tank	8 h	Ops
10126-2	Confined Space Entry of Fuel Oil Tank	1 h	Ops, Labr, Mech
10126-3	Open Fuel Oil Tank	2 h	Mech, Labr
10126-4	Vacuum Tank of Sludge	4 h	Labr, Honey Truckin'
10126-5	Close Up Fuel Tank	2 h	Mech, Labr
10137	Rebuild Fuel Control Valve	3 h	Mech
	Auxiliary Equipment Work		
	Rebuild P-1A		
10102-1	Lock-Out, Tag-Out of P-1A	0.5 h	Ops, Mech, I/E
10102-2	Remove P-1A— Bring to Shop	2 h	Mech[2], C/Pkr
10102-3	Overhaul P-1A	12 h	Mech
10102-4	Set P-1A on Base and Install	2 h	Mech[2], C/Pkr
10102-5	Align P-1A	2 h	Mech, Laser
	Rebuild P-1B		
10103-1	Lock-Out, Tag-Out of P-1B	0.5 h	Ops, Mech
10103-2	Remove P-1B—Bring to Shop	2 h	Mech[2],C/Pkr
10103-3	Overhaul P-1B	12 h	Mech
10103-4	Set P-1B on Base and Install	2 h	Mech[2], C/Pkr

(*continued*)

Table A-1 (continued)

Work Order #	Task	Duration	Resource
10103-5	Align P-1B	2 h	Mech, Laser
	Deaerator Work		
	Drain Deaerator	1 h	Ops
10101-1	Open Deareator	1 h	Mech
10101-2	Wash Deareator Internals	1 h	Labr
10101-3	Confined Space Entry of Deareator	1 h	Ops, Mech
10101-4	Check Deareator Thickness	16 h	Mech[2]
10101-5	Close Up Deareator	1 h	Mech
	Misc. Mechanical Work		
10127	Rep. Insul.—Return Line—PO# 5533	72 h	Thermal Experts
10128-1	Rem. Insul.—Proc. Feed—PO# 5533	12 h	Thermal Experts
10128-2	Repair Steam Tracing—Process	16 h	Mech[2]
10128-3	Inst. Insul.—Proc. Feed—PO# 5533	24 h	Thermal Experts
10129	Rebuild Control Valve—Condensate	3 h	Mech
10130	Change All Pressure Gauges	6 h	Mech
10131	Rebuild Deion H_2O Valve	3 h	Mech
10132	Fix Dent in Stack— PO# 5531	24 h	Acme Boiler
5937	Change Gasket in Main Header	3 h	Mech[2]
10201	Drain and Clean Chem Treatment Syst	3 h	Ops, Labr
10135	Open and Clean 25 Steam Valve Stems	12.5 h	Mech
10136	Change Zeolite in Softener	16 h	Mech[2]

(continued)

Table A-1 (*continued*)

Work Order #	Task	Duration	Resource
10119	Rebuild BFW Control Valve	3 h	Mech
10121	Clean Firebox Sight Glasses	1 h	Mech
	Misc. Instrument and Electrical Work		
10112	Overcurrent Test on O/Ls and CBs	6 h	I/E
10113	Resistance Test of Starter and Feeders	6 h	I/E
10123	Tighten Connection— BL-1 Starter	0.5 h	I/E
10134	Change Breaker on P-3B	2 h	I/E
10114	Calibrate Ratio Controller—Boiler	1 h	I/E
10115	Calibrate T-1 Level Control	1 h	I/E
10116	Verify Boiler Shutdown Systems	3 h	Ops, I/E
10124	Overhaul Flame Sensor and Control	4 h	I/E
10109	Calibrate TCs on Boiler to Stack	3 h	I/E
10107	Clean Electrodes, Calibrate LC-1	2 h	I/E
10108	Clean Sensor, Calibrate O_2 Analyzer	2 h	I/E

Appendix B

Identifying Electrical Work to Be Performed During a Shutdown

When a shutdown or turnaround is planned for industrial facilities, a plan for electrical maintenance that requires an electrical outage is usually not included. This is unfortunate because electrical outages provide the planner with a unique opportunity to define future required work. This appendix includes a partial list of work performed before and during an electrical outage.

Table B-1 describes tasks that should be performed before an outage, and Table B-2 describes tasks during an outage.

Electrical Shutdown Checklist

The remainder of this appendix provides a checklist of information to be collected for an electrical shutdown. The information is broken down into two major milestones:

- Months and weeks prior to the shutdown
- During the shutdown

Months and Weeks Prior to the Shutdown

The following list shows all work that should be investigated (in the order of priority) before any plant shutdown is to occur:

- Poles and cable support structures
- M/V breakers + fuse disconnects
- L/V switchgear (metal clad)
- Transformer checks
- Infrared scan
- Motor starters and MCCs
- General shutdown checklist

The following sections show information that should be gathered at each stage of the shutdown.

I. Poles and Cable Support Structures

General Walking the system periodically is the key to inspections (use binoculars where necessary). Aerial cable installations should be inspected for mechanical damage caused by vibration, deteriorating supports, or suspension systems.

Table B-1   Before the Outage

Step	Task	Description
1.	Conduct an Infrared Scan	A standard three-phase power system will usually have equal current flowing in all three phases. As a result, all heat given off by the cables, contacts, and connections should be equal in adjacent phases. If a connection is loose, corroded, or dirty, it will heat up more than the other phases. The infrared scanner provides a method for finding these hot spots that, if left uncorrected, could result in catastrophic failure.
		If a possible problem is identified with the scan, an infrared thermometer is used to determine the severity. A 5°C increase above normal operation is considered to be a point of concern.
2.	Transformer Oil Testing	Transformer oil samples can be taken from most oil-filled transformers while they are running. Two types of tests should be performed on the oil: basic screening and dissolved gas analysis.
		Basic screening tests determine the serviceability of the oil as a cooling medium and as an insulator. Dielectric, interfacial tension, and acidity are some of the tests performed.
		Dissolved gas analysis (or gas-in-oil) can reveal something about the condition of the transformer windings, internal connections, and iron core. For example, the presence of combustible gases (such as carbon monoxide) indicates destruction of the winding-phase separators. Acetylene, in even very small quantities, can indicate a high-energy arc in the transformer. When high combustibles are found, retests are usually performed to see if the gases are increasing.

Detail Note all discrepancies for correction during a future shutdown.

1. Infrared Scan to be performed on all cables and connections. See the Infrared Scan section for details.

2. Inspect static lines and guide wires for corrosion, abrasion, tension, and tight connections to supports.

3. Inspect poles and cross-beams for structural integrity. Identify broken insulators.

4. Potheads should be inspected for oil or compound leaks.

Table B-2 During the Outage

Step	Task	Description
1.	Overcurrent Tests on Molded Case Breakers, Thermal Overload Protective Devices, and Other Protective Devices	Overcurrent devices have published *time vs. current characteristics*. Trip tests are done to determine if a protective device is working within its performance rating.
2.	Megohm Testing of Rotating Machines, Transformers, Insulator Bushings, and Power Cables	The megohmmeter is an instrument designed specifically to measure insulation resistance directly in megohms (one megohm equals 1,000,000 ohms). Insulation resistance can be measured without damaging the insulation and furnishes a highly useful guide for determining the general condition of insulation. Megohmmeter testing may be used to test the insulation resistance between conductors of separate circuits or between the conductors and ground.
		The resistance measured in *good insulation* will begin very low and then begins to level off after about 10 to 15 minutes. If the insulation is *wet* or *dirty*, the resistance will level off quickly.
		There are two insulation values that can be used to ascertain the quality of the insulation: the spot reading and the polarization index.
		The *spot reading* is the megohm reading after one minute. The IEEE Standard 43 establishes a limit for this spot reading as the Kvolt value of the device under test plus one (in megohms). In other words, for a 460 volt motor, the insulation resistance limit would be 1.46 megohms. This value may be acceptable as a *troubleshooting limit* but would prove to be unacceptable if extended service is to be ensured.

(continued)

Table B-2 (*continued*)

Step	Task	Description
		Insulation measuring 1.46 megohms may last 6 months to 1 year, or it could fail in a day.
		A better limit is 100 megohms. New insulation has a megohm reading close to 2000 megohms. When this value degrades to 100 megohms, there is still time to repair the damage. If it is allowed to degrade further, to 1.46 megohms, it is close to failure.
		Another test measures the dielectric absorption. A dielectric absorption test is an extension of insulation resistance testing for longer than the conventional 1-minute period. To perform this test, a motor-driven or battery-operated megohmmeter is required. In this test, the insulation resistance is recorded at 1 minute and 10 minutes when readings plateau. Clean and dry insulation in good condition will steadily increase in resistance value over the 10-minute period. Dirty or moist insulation will plateau quickly and at a relatively low value of resistance.
		A number called the polarization index provides an easy way to evaluate the results. The *polarization index* is the ratio of the 10-minute reading to the 1-minute reading, calculated as shown here:

$$\text{Polarization index (PI)} = \frac{\text{R10 (Resistance at 10 minutes)}}{\text{R1 (Resistance at 1 minute)}}$$

		If the insulation resistance reading *more than doubles* itself from 1 minute to 10 minutes, then the insulation is considered to be in good condition.
		If a motor has a poor polarization index, it usually does not require a rewind. In most cases the motor can be dried out and varnished, resulting in a much-improved megohm spot reading and polarization index.

(*continued*)

Table B-2   (continued)

Step	Task	Description
3.	DC High Potential Testing	The dielectric strength of an insulation system determines the level of voltage a piece of equipment can withstand without arcing over to another phase or to ground. It also determines how much overpotential (such as a voltage surge) the insulation can withstand. Dielectric strength testing is accomplished by placing a calculated voltage on conductors to test their insulation. This may be either an AC voltage using an AC high potential (hipot) tester or a DC voltage using a DC hipot tester. DC high potential testing is preferred for maintenance use because, if performed properly, it can be nondestructive. AC high potential testing is a more sensitive procedure and the possibility of a poor device or cable failing under the test is much greater.
4.	Mechanically Operate and Inspect Large Breakers	Large circuit breakers should always be removed and inspected during electrical outages.
5.	Clean Switch Gear and Other Cubicles	Switch gear and all other cubicles should be thoroughly cleaned during an outage. Debris left behind might eventually lead to a major problem.

II. M/V Breakers + Fuse Disconnects

1. Inspect enclosure for dust, dirt, rodents, reptiles, corrosion, and tracking.

2. Record the number of operations (if counter is present) and check indicating lights.

3. Reset any tripped relay flags. Investigate reason for flag and determine if a long-term corrective action is necessary.

4. Operate space heaters. Use amp probe to determine if they are operating properly.

5. Inspect batteries for proper water level and low specific gravity. Clean as necessary.

6. Contract for overcurrent and other relay testing (annually).

III. L/V Switchgear (Metal Clad)
Detail

1. Test voltage with DVM, phase to phase, and phase to neutral to identify any grounds. Record all readings. Verify ground lights are operating. Voltages should be within 1 percent of each other from phase to phase. All readings should be ±2 percent of transformer nameplate rating (that is, 480 volts ±9.6 volts). Investigate any irregularities.
Actual Voltages:
L/V Switchgear Name: _____

A to B _____ volts	A to G _____ volts	
B to C _____ volts	B to G _____ volts	
C to A _____ volts	C to G _____ volts	

2. Infrared scan for loose or corroded connections. See the infrared scan section.

3. Contract for overcurrent tests to be performed on breakers (at least every three years). Be sure spare breakers are available before the shutdown in case one needs to be replaced.

IV. Transformer Checks
Detail

1. Record primary and secondary volts and amps. Verify that current is below transformer ratings on all phases. If current exceeds self-cooled or partially force-cooled levels, be sure proper cooling equipment (fans and/or pumps) is operating to dissipate heat.

   Voltages should be measured either with existing meters or a DVM. Phase to phase or phase to ground readings are both acceptable. Readings must be between ±5 percent of rated voltage (tap setting value). Abnormal voltages indicate possible supply voltage changes and the utility should be contacted. Three-phase voltage unbalance is also a possible problem and abnormal changes should be investigated. Unbalance of 1 percent or greater use is severely detrimental to electric motors.

   Actual Voltages:
   Transformer Name: _____

A to B _____ volts	A to G _____ volts	
B to C _____ volts	B to G _____ volts	
C to A _____ volts	C to G _____ volts	

2. Record hot spot and top oil temperatures. (Hot spot ther-mometers are usually found on large transformers only.) At peak loads, the transformer hot spot temperature should be on the high end of its normal operating range. Hot spot temperature at low loads will be lower than at peak. Top oil temperatures may differ only slightly from inspection to inspection (because of the forced cooling system). Any major change in top oil temperature is an indication of either cooling fan or pump failure, improper fan and pump control settings, or major failure of the transformer oil.

Temperatures at 40°C Ambient, Upper Limit Is

	Gauge	Drag Needle* (High-point)
Hot Spot	65°C	No greater than 105°C
Top Oil	55°C	No greater than 95°C
(on most transformers)		

*The drag needle is the needle on the gauge that is pushed to the highest transformer temperature by the indicating needle. This needle should always be reset to the indicating needle position after inspection.

Actual Transformer Gauges

Transformer Name: _____

	Gauge	**Drag**
Hot Spot	_____ °C	_____ °C
Top Oil	_____ °C	_____ °C

3. Record oil level. Oil levels will be slightly lower at low load periods and on cold days. Be sure to mark this lower limit on the level gauge as a guide to any loss of oil in the transformer.

   Sight glass type gauges should be full to a level just below the transformer tank top. Magnetic type gauges will have a LO range marked on them.

Actual Oil Level

Transformer Name:_____

Level Gauge _____ (HI, LO, NORM, MID, and so on)

4. Nitrogen blanket system should be checked for maximum and minimum pressures (for maintained blanket only).
   Limits Blanket Pressure;
   +8 psi Gage Maximum to +1 psi Gage Minimum

Nitrogen Tank Pressure

150 psi Minimum

Any pressure outside the limit should be investigated immediately.

5. Inspect for physical damage. Note any dents, scratches, loss of paint, and corrosion on the transformer and associated equipment. Inspect transformer mountings for sagging or deterioration.

6. Note any leaks from transformer tank, fittings, cooling tubes, and bushings. Some leaking of transformer oil is common during initial start-up only. Any leaking component should be immediately cleaned and the condition should be monitored. If leaking persists, the transformer must be deenergized and the source of the leak corrected.

7. Verify proper auxiliary device operation. Operate (on manual) all pumps and fans. Note proper operation of flow indicators (for pumps). Identify any excessive vibrating device. Shut down all fans, inspect and clean dirty fan blades.

Caution

Be sure temperature limits are not exceeded while fans are down.

8. Take vibration tests (in in/sec velocity) on transformer and associated equipment. The transformer tank readings should be consistent with previous readings while noting transformer loading. (Mark a spot on the transformer for consistent readings.) A 20 percent increase above initial (new) readings of vibration are an indication of mechanical looseness in the transformer or core failure.

   The motors, pumps, and fans *must not* have vibration levels of 0.3 in/sec or greater. These readings should be taken at the equipment bearing housing.

9. Oil and operate any and all locks on the transformer, including manual tap changer locks. Do not operate tap changer.

10. Inspect and tighten external ground connections.

11. Perform infrared scan of transformer. (See the Infrared Scan section.) Identify any loose or corroded connections, overheated bushings, or blocked cooling fins. Over temperature bushings or connections are detected by uneven heating on the three-phase transformer connections. Blocked cooling fins are identified by a cooler (or dark in infrared) fin in a group of fins.

12. Perform basic oil screening and dissolved gas analysis on all plant transformers. Contract this service through Specification for Transformer Insulating Oil Maintenance Testing.

13. Contract repair service if required.

V. Infrared Scan

General An infrared study is performed for the purpose of detecting loose or corroded connections on electrical equipment. Electrical current flowing through copper or aluminum conductors generates a measurable amount of heat. This heat is mainly caused by the resistance to current flow that exists in all metal conductors. The heat generated, or Power Lost, as it relates to resistance (R) and current flow (I), is described as follows:

$$\text{Power Lost} = I^2 \times R$$

If a loose or corroded connection should occur, the resistance to current flow increases greatly. So does the generated heat. This heat may be enough to distort and melt the connection, resulting in an electrical outage. This work is usually performed by an outside service because of the expense of the equipment required. An infrared imager with some sort of recording (such as a photo) and a temperature detector are required for this survey.

Detail

1. Conduct an inspection and document all equipment and locations where an infrared scan can be performed. Items usually considered are the following:
   a. Aerial cable
   b. All cable connections
   c. Bus work and connections
   d. L/V substations
   e. Motor control centers
   f. Motor starters and C/B connections
   g. Transformer cooling fins

2. Install hinges or handles on equipment, which may be difficult to access while hot. Often the installations of handles on large switchgear enclosures make access quicker and safer during infrared inspections.

3. Schedule the infrared scan for a period in which load in the plant is near maximum. This will allow the worst problems to better exhibit themselves.

4. Perform infrared scan. Abnormal heat levels can more easily be detected in three-phase power systems. Most three-phase power connections will have an equal amount of electrical current flowing through each phase (hence, an equal amount of heating on all connections). If a loose or dirty connection should occur, the same current would flow through all three phases, but more heat will be generated at the loose connection (because of increased resistance, or –R). Eventually, unbalanced voltages or even single phasing may occur, causing failure not only at the faulty connection but also to all other equipment connected to it.

The infrared viewer can detect any unbalanced heating in a three-phase load, and using the measured temperature, the severity of the problem can be deduced. All three-phase electrical connections and cables should be at a temperature within 5°C of each other.

The following items must be recorded on problems found through the infrared scan:

a. Ambient temperature of the equipment in the area

b. Temperature of all three phases (if a three-phase system)

c. Current through the connections or cable

d. A drawing or sketch of the equipment with temperatures identified

e. Date when problem was found

f. Recommended latest repair date

g. Take photos of all problems

VI. Motor Starters and MCCs
Detail

1. Inspect all starter cubicles in the area to be shut down for the following:

a. Dirt, dust, rodents, spider webs, and so on

b. Corrosion on starter components or cubicle

c. Noisy coils or contacts

d. Burnt or distorted current carriers

2. Perform an infrared scan on all operating starters. (See the Infrared Scan section.)

3. Make a list of all associated motors, and record all nameplate data.

4. Make a list of all parts and equipment that will be required for the test.

5. Inspect all junction boxes and cable trays.

VII. General Shutdown Checklist (Suggested List)

1. Where will temporary power be needed?

2. Will emergency generators or temporary generators be needed?

3. What special tools/equipment will be needed?

   a. Compressors

   b. Temporary generators

   c. Scaffolding

   d. Lifts

   e. Cranes

   f. Special instruments

4. What contractors or consultants need to be called in?

5. What spare parts or supplies need to be purchased (or check stores)?

6. List all work orders to be performed during this shutdown.

7. List all P.O. numbers to be issued for the shutdown.

8. Make a plan. What needs to be done first? What jobs must you coordinate with other crafts or production? In what sequence should the work be done?

9. Make sure everything (instruments, equipment, and so on) is in working order prior to shutdown.

During the Shutdown

With equipment ready for maintenance or deenergized, be sure to gather data for the next shutdown and identify parts and materials needed.

I. L/V Breakers (Metal Clad)

General L/V Breakers should be removed from service and tested at the designated interval or after the breaker has opened to interrupt a damaging fault. Stored energy should be discharged prior to any maintenance.

Detail All performed annually.

1. Remove arc chutes. Inspect, adjust, and clean as necessary for broken parts, missing arc splitters, metal splatter, and burning

on interior surfaces. Snuffer screen must be clean. Replace arc chutes with new ones, if required.

2. Inspect main contacts for pitting, spring pressure, overheating signs, alignment, over-travel, or wipe, evidenced by slight impressions on the contact surfaces. Clean or replace as necessary. Silver contacts will show more discoloration and indentation than copper. This is usually acceptable. Never file contacts. Nothing more abrasive than crocus cloth or Scotch Brite should be used.

3. Inspect arcing contacts for alignment, overtravel, or wipe (evidenced by slight impressions on the contact surfaces) and for arc erosion. Arcing contacts are the last to open during a fault. Verify this by slowly manipulating the operating mechanism. Replace as necessary.

NOTE

Contacts should be replaced in threes.

4. Disconnect finger clusters. Inspect for proper adjustment and spring pressure.

5. Structure or frame to be checked for proper alignment and loose or broken parts.

6. All insulating materials to be inspected for cracks, breaks, or signs of overheating.

7. Oil all parts such as bearings, racking screws, and sliding screws. Do not apply lubricant to contacts.

NOTE

The following tests should be performed by competent personnel. Test equipment required to achieve reasonable test currents may dictate the use of outside services. (Greater than 30,000 amps of continuous test current is sometimes required.)

8. Contact resistance test (breaker closed):
   Use a Ductor, Wheatstone Bridge, or other microohmmeter to measure contact resistance on all three phases (from input to output). Resistance shall not be more than 2000 microohms, and values should not deviate from pole to pole, or from similar breakers, by more than 50 percent. Record all readings.

9. Insulation resistance test (breaker open):
   Use a 1500-volt megohmmeter to test each phase (on the load and line side) to ground (case), phase to phase, and

phase line to phase load on each phase. Readings must not be below 1000 megohms. All readings to be recorded at 60 seconds.

10. Overcurrent trip testing (electromechanical and solid-state type).

    Using high-current test device, verify trip curves by testing electromechanical trip devices at 150 percent, 300 percent of rated current, and verify instantaneous trip setting. Test current to be injected at the contact fingers with proper size stab. Verify trip curves by testing solid-state trip devices. All existing trip settings to be verified.

NOTE

These tests, inspections, and adjustments should be performed in conjunction with L/V Breaker tests. Verify that no parts of the power or control circuitry are energized by back feed from other power or control sources.

11. Completely clean with vacuum and lightly dampened rags.

12. Clean and repaint enclosure as necessary. Clean all vents. Oil hinges.

13. Check bolts, bus connections, and splices for required tightness. When in doubt, use the standard torque table.

14. Clean all insulators and insulating material and inspect for cracks.

15. Clean and inspect breaker stabs for pitting and effects of overheating.

16. Inspect cables and connections for evidence of overheating or frayed insulation.

17. Clean, align (as necessary), and oil racking mechanisms.

18. Test all meters for accuracy. Repair or replace as necessary.

19. Verify all grounding is intact.

20. Test conductivity of all aluminum cable and bus connections. Use a Ductor, Wheatstone Bridge, or other microohmmeter to measure connection resistance. Resistance must be below 2000 microohms. Use Belleville washers when bolting aluminum cable lugs to equipment.

21. Use a 500-volt megohmmeter to test each phase on the main bus from phase to ground and from phase to phase. Measure resistance on all incoming and outgoing power cables. All

readings are to be taken for 60 seconds. All readings must be above 1000 megohms.

22. Reinstall breakers and verify the disconnect finger clusters are making proper contact with stabs. Do not lubricate contacts. Measure contact resistance when proper contact is in doubt.

II. Low- and Medium-Voltage Motor Starter Test and Maintenance
(See data collection sheet.)
Detail For low-voltage starters only:

NOTE

Be sure the motor is down and power to the breaker is off.

1. Clean and vacuum all cubicles. Oil all hinges. Do not lube starter components.

2. Disconnect the line and load leads from the breaker and perform the following megohm tests. All readings must be above 500 megohms.
   Line side:

   > Phase to ground, A phase, with B and shorted to ground

   > Phase to ground, B phase, with A and C shorted to ground

   > Phase to ground, C phase, with A and B shorted to ground

   Load side:

   > Phase to ground, A phase, with B and shorted to ground

   > Phase to ground, B phase, with A and C shorted to ground

   > Phase to ground, C phase, with A and B shorted to ground

   Other:

   > A line phase to A load phase

   > B line phase to B load phase

   > C line phase to C load phase

   All megohm tests are to be performed for 60 seconds unless megohm reading is greater than 2000 megohms, then motor is good.

3. Remove the breaker and have it tested (by vendor)

NOTE

The tests in (4) and (5) are performed at three years. These tests are best conducted by an outside service. (Special test equipment is required to provide currents in excess of 30,000 amps.)

4. Using a Wheatstone Bridge, Ductor, or other low-ohm device, measure contact resistance of all breakers. Resistance shall not be higher than 2,000 micro ohms.

5. Overcurrent test all molded case circuit breakers and compare results with timecurrent characteristics. Emphasis must be put on instantaneous trip operation. If the breaker does not trip on first test, raise test current above trip setting by 50 percent (but not to exceed breaker interrupting rating) and test again until unit trips. Test all three phases, recording as-found and as-left readings. Replace any breaker that fails to trip or trips earlier than design characteristics.

Detail For low- and medium-voltage starters:

6. Meg out the starter as described for the breaker in step (2). All readings should be above 500 megohms (taken for 60 seconds unless it is above 2000 megohms). If the starter megs bad, either remove and clean it or replace it.

7. Inspect contacts if accessible. Clean or replace as necessary. Never file or use sandpaper on contacts.

8. Check arc hoods for cracks, breaks, or deep erosion.

9. Change cracked or embrittled coils.

10. Reinstall all starter cubicle parts and tighten all connections.

11. Paint cubicle if necessary.

III. Transformers (Oil Filled)
General All tests, repairs, and inspections in this section must be made after the transformer is completely discharged. All instrument transformers must be disabled either by removing fuses or shorting the secondary and primary windings.

Before work is performed in the transformer, all tools should be counted and listed. The list should be checked after the end of work to make sure no tool has been left in the transformer. All tools used *in the transformer* should be tied to a long string connected *outside the transformer*. Any person entering the tank should remove any loose items (change, keys, and so on) from pockets prior to entering. A transformer tank is just that—a tank—and a standard tank entry permit must be completed before anyone is allowed to enter.

Some tests, repairs, and inspections may not apply to all transformers.
Detail Annually

1. Perform required maintenance on the transformer as determined by inspections.

2. Contract oil servicing, as required.

3. Inspect pressure-relief diaphragm for cracks or holes and verify proper operation. (Possible cause of pressure in sealed type transformers remaining at zero.)

4. Clean bushings and lightning arrestors and inspect surfaces. Check level and refill as necessary all oil-filled bushings. (Westinghouse type 0 bushings are oil filled.) Check for leaks, hardened bushing gaskets, and corroded or broken ground connections.

5. Inspect tap changer, follow manufacturer's instructions, and contact replacement frequency requirements.

6. Paint tank as required. Wire-brush rust spots and prime paint. *Never paint fan blades*.

7. Make undercover inspection. (Positive pressure should be applied to prevent moisture entrance.) Note moisture or rust under the man-way covers at the top of the tank. Also, note oil sludge deposits, loose bracing, or loose connections. Take corrective action as necessary. (Moisture or rust on the cover may be an indication of water in the oil. Oil screening test should be reviewed.)

8. Check all ground system connections, inside and outside the transformer, for corrosion. Correct as necessary.

9. Check *all* drain valves for proper operation. Check, tighten, or replace gages, valves, and fittings.

10. Check tightness of all electrical connections on transformer.

11. Insulation Resistance (do not perform this test if the transformer is emptied of oil). Using 500, 5000-volt megohmmeter conduct time-resistance tests for insulation resistance. Prior to test, note insulation temperature (with either infrared thermometer or resistance method—see Item 12). All spot readings should be corrected to 40°C base.

12. Test should be performed on all windings to ground and between windings. Guard circuits must be used to drain unwanted leakage.

    The following describes all possible combinations on a three winding transformer that should be checked:

    a. High to Low, Tertiary and Ground @ High Voltage

    b. High to Ground @ High Voltage

    c. Low to Ground @ Low Voltage, then High if it passes

d. High to Tertiary @ High Voltage

e. Low to Tertiary @ Low Voltage, then High if it passes

Permanently connected windings should be considered as one winding. All tests are to be conducted for a minimum of 10 minutes, noting insulation resistance at 30-second intervals.

Limits

$$PI = \frac{R\ 10\ minutes}{R\ 60\ seconds} = \text{should be greater than } 2.0$$

Resistance @ 60 seconds (corrected to 40°C) should be equal to or greater than 500 megohms.

Correcting Insulation Resistance for Temperature

$$Rc = Rm \times KI$$

where the following is true:

Rc = Corrected insulation resistance

Rm = Resistance measured after 60 seconds

KI = Correction factor from the following table

°C (Winding temperature)	KI	°C (Winding temperature)	KI
0	0.06	45	1.41
5	0.09	50	2.00
10	0.13	55	2.82
15	0.18	60	4.00
20	0.25	65	5.64
25	0.35	70	8.00
30	0.50	75	11.28
35	0.71	80	16.00
40	1.00		

13. Low Resistance Winding Test.

Using a Ductor, Wheatstone Bridge, or other low-ohm (microhm) meter, measure and record transformer winding resistance. All high-voltage windings should read the same, as should all the low-voltage windings. All high-side winding resistance should be within +5 percent, as should all low-side windings. If the transformer has cooled to ambient air temperature at the time of the test, record the temperature indicated. Correct all resistance readings to 40°C by using published tables for copper or aluminum windings.

Rs = Kw × Rm

where the following is true:

Rs = resistance at desired temperature (40°C)

Rm = measured resistance

Tm = temperature at which resistance was measured

14. Turns Ratio Test:

This test is best performed by a qualified electrical testing contractor. The results of this test should verify the turns ratio of all transformer windings (that is, primary voltage divided by secondary voltage). Also, all transformer winding ratios should be within +0.5 percent of calculated ratio.

Turns Ratio

Test results can be used in conjunction with winding resistance measurements to diagnose transformer problems. A positive turns ratio test with a negative winding resistance test indicates loose connections in the transformer, possibly in the tap changes.

15. Insulation power factor test (optional for transformers less than 2000 KVA).

This test is best performed by a qualified electrical testing contractor. The results of this test should be below limits established by Doble Engineering, which are the following:

Below .5 percent for a new transformer

Below 1 percent for a transformer that has been in service for 10 years or more

16. Exciting Current Test (optional for transformers less than 2000 KVA).

This test is best performed by qualified electrical testing contractors. The pattern of test data should be looked for here. Two similar current readings for the outside phases and a lower current reading for the center phase of the transformer are indicative of good readings. Shifting of the transformer core laminations will result in a major change from this pattern or from previous readings.

Appendix C
Machinery Inspection Prior to and During a Shutdown

The shutdown process provides an opportunity to build reliability back into equipment. True reliability can only be achieved if high standards of inspection and workmanship are attained. These standards must be in writing and well understood by all those involved. Data collection before and during the shutdown is key to building this reliability. Inspections conducted before a shutdown help define the scope and duration of work that may be performed. Details for current and future shutdowns can also be obtained once the equipment is shut down and opened.

This appendix provides standards and inspection routines for common machinery and components to help ensure that the coming shutdown (and future shutdowns) provides the most reliable result.

Bearing Data

One key to job planning is the existence of a detailed spare parts list. Running out of parts (or rushing in shipments in the middle of a job) is a major disruption during a shutdown. The extra delivery time can extend the job duration and may affect the start of subsequent jobs, hence the entire shutdown. This presents a particular dilemma for the planner when equipment manuals or detailed spare parts lists (bills of material) do not exist.

The nameplate on some equipment may provide an answer. Many manufacturers identify bearing information using a standardized bearing designation defined by the American Bearing Manufacturers Association (ABMA):

> American Bearing Manufacturers Association
> 2025 M Street, NW, Suite 800
> Washington, DC 20036
> Phone: 202-367-1155
> Fax: 202-367-2155
> Web: http://www.abma-dc.org

The ABMA assigns a generic identifier to bearings manufactured around the world, as opposed to the manufacturer's part number. The ABMA standard for designating bearings uses alternating sets of numbers and letters, which represent different components and design characteristics of the bearing. This identifier is known to all bearing distributors and is available in numerous interchange references.

Figure C-1 explains this system and includes the specific information encountered on many alternating current (AC) motor nameplates. A complete copy of this standard can be obtained from the ABMA.

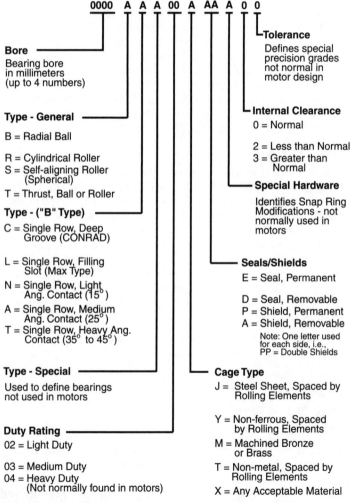

Figure C-I ABMA nomenclature for motor nameplates.

A couple of examples of how this chart could be used may be helpful.

The following are bearing numbers off the nameplate for a vertical housing motor:

Lower Bearing—70BC03JPP3, Upper Bearing—110BT02M

The following information can be derived using the data in Figure C-1.

70BC03JPP3

70 = 70 mm Bore

BC = Radial Ball, Single Row, Deep Groove

03 = Medium Duty

J   = Steel Cage

PP = Double Shield

3   = C3, larger than normal internal clearance

An equivalent SKF bearing is a 6308 2Z C3 or EM.

110BT02M

110 = 110 mm Bore

BT = Radial Ball, Single Row, heavy Angular Contact

02 = Light Duty

M   = Bronze Cage

An equivalent SKF bearing is a 7222.

Shaft Alignment Standards

Sometimes, equipment that is removed from service for repair or inspection is returned to service in worse condition. Managers fear this scenario, and they may suggest the adage, "If it isn't broke, don't fix it." Poor machinery alignment, between driver and driven equipment, is a common mistake maintenance workers make during the frenzy associated with a maintenance shutdown.

Proper shaft alignment is achieved when the two centerlines of coupled machinery are the same or collinear. Shaft alignment is often referred to as coupling alignment because the required measurements are commonly taken on the coupling rim or face. There are a number of methods used to achieve proper alignment:

- *Straightedge*—Used mostly for rough alignment.
- *Feeler Gauge*—Can be used for alignment in combination with gauge blocks.

- *Taper Gauge*—Used for achieving angular alignment measurements on close-coupled machinery.
- *Face-Face Distance*—This method is used primarily in cooling tower fan alignment and other machinery with long coupling inserts.
- *Rim-Face Dial Indicating Method*—Uses dial indicators to determine offset and angular misalignment. The name is derived from the fact that readings must be taken on the face and rim.
- *Reverse Dial Indicating Method*—Growing more popular, the reverse dial method provides the most accurate results under common conditions in chemical and petrochemical machinery. This method is the basis for many computer-aided alignment systems.
- *Laser Alignment Method*—Potentially the most accurate method of alignment, but does not preclude the use of dial indicators for run-out readings.

Shaft alignment methods attempt to correct parallel misalignment (offset) and angular misalignment, as illustrated in Figure C-2.

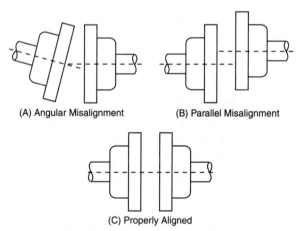

(A) Angular Misalignment   (B) Parallel Misalignment

(C) Properly Aligned

Figure C-2 Parallel and angular misalignment.

Angular misalignment is a condition where two shaft centerlines are at an angle to each other. *Parallel misalignment* is a condition where no angular misalignment exists and the shafts are parallel but offset from each other. Most equipment misalignment problems will be made up of a combination angular and parallel misalignment.

One point should be made before the discussion about alignment continues. The goal of shaft alignment is to line up the centerlines of two machines and make essentially the same centerline. Often, leveling of the machinery is confused with alignment of the machinery. The fact is that two shafts can have the same centerlines, even though they are not level.

Run-Out Measurements

Run-out measurements should be made as a preliminary check of the coupling, as well as the driver and driven equipment shafts. Most coupling manufacturers have quality-control programs to ensure a bore that is on-center. However, some coupling halves (or hubs) pass by quality control with an off-center bore.

To check for coupling run-out, first mount both coupling halves on the driver and driven shafts. The coupling halves should slide on the shaft, but should not wobble once in place. If a wobble exists, the coupling half may be bored too large, or the shaft is worn or undersize. If this wobble is not corrected prior to tightening the setscrews, an excessive run-out reading may be detected in the next step. Measure and compare the coupling and shaft diameters with a micrometer to isolate the problem. The coupling bore should not be more than 0 to .002 of an inch larger than the shaft.

Coupling run-out is checked by mounting a dial indicator on an immovable surface (such as the machinery base) and depressing the tip of the indicator against the coupling rim. A magnetic base clamp is best for this purpose.

With the indicator needle zeroed, the shaft is slowly turned. Run-out is measured by turning the motor shaft 360 degrees while noting the total indicator travel, both in the positive and negative direction. The total positive and negative movement is called the total indicator reading (TIR).

Make sure the indicator tip is perpendicular to the coupling rim surface. If the tip of the indicator is at an angle other than 90 degrees to the surface, the resultant readings will be less than the actual run-out.

A problem exists if the TIR is greater than .005 inch. The cause of the problem is either a bent shaft or bad coupling. To isolate the problem, remove the coupling half and repeat the run-out check on the shaft.

The shaft is turned 360 degrees while noting the indicator needle travel in both the positive and negative direction. If the TIR is greater than .005 inch, the shaft is bent and must be repaired or replaced prior to attempting an alignment procedure. If the TIR is less than

.005 inch, the coupling half may be bored off center and should be replaced.

Alignment Data Conventions and Checks

Alignment readings are usually represented on a circle. The position of the indicator is shown in the center of the circle. This is usually represented as M-P (indicating that the readings are being taken from the motor to the pump) or P-M (indicating that the readings are being taken from the pump to the motor). Rim readings are recorded on the outside of the circle and face readings are recorded on the inside of the circle (Figure C-3).

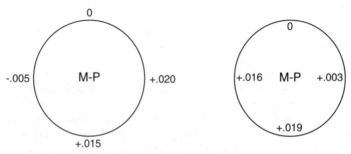

Figure C-3   Recording alignment data.

Note that when the top and the bottom readings are added together, they equal the algebraic sum of the left and right readings:

$$0 + .015 = .020 - .005$$

The same is true for the face readings:

$$0 + .019 = .016 + .003$$

This is a quick check of the consistency of the dial readings.

The data positions on the circle are often referred to as 0 degrees, 90 degrees, 180 degrees, and 270 degrees readings. Another convention in use refers to hour positions on a clock face. The corresponding data positions are then referred to as the 12 o'clock, 3 o'clock, 6 o'clock, and 9 o'clock readings. Figure C-4 shows some examples.

The Definition of a Line

One rule of geometry is that a straight line is defined as the shortest distance between two points. A corollary to this rule is that a line can be defined by just two points. The reverse dial shaft alignment

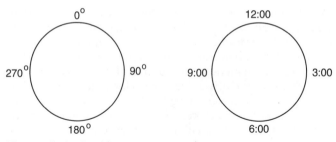

Figure C-4   Data conventions.

method is a derivative of this fact. The position of one shaft relative to another is determined by identifying two points on that shaft centerline. Figure C-5 shows how this is done.

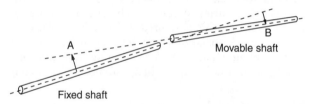

Figure C-5   Alignment of two shaft centerlines.

Point A is determined by locating the position of the movable shaft in relation to a point on the fixed shaft. Point B is determined by locating a fixed point on the fixed shaft to a point on the movable shaft. Together, points A and B locate the position of the movable shaft in relation to the fixed shaft.

As stated earlier, the accuracy of the indicator data can be checked by comparing the sum of the top and bottom readings to the sum of the left and right readings. These sums should be equal or not differ by more than .001. If the difference of the sums exceeds .002 or more, the readings must be retaken. Erratic or erroneous data may be the fault of a loose indicator setup or a faulty indicator.

Large motors may provide a minor difficulty when collecting dial indicator (or laser) alignment readings. The rotor of an electric motor runs in a magnetic field set up by the stator. Most large electric motors have sleeve bearings so that the shaft and rotor are free to move inside the bearing to seek its steady-state magnetic center. This center can vary with load. Large motor shafts are marked to show the magnetic center, usually by punch marks or an etched line.

The mark will move to the edge of the inboard seal when the motor is running at full load. Motors with no markings can be assumed to have a magnetic center at the center of the total endplay.

The mechanic who aligns the motor should make sure it will be free to move to its magnetic center, unbound by the coupling or by the axial thrust of the driven machine. This is accomplished by setting the gap between the coupling halves such that it allows the motor to seek its magnetic center. With the magnetic center and endplay limits marked with a colored marker, observe the motor during start-up and partial load to ensure that the shaft is free to move. If the motor is forced off its magnetic center by the coupling, vibration and bearing wear will increase.

Large electric motors with sleeve bearings are usually coupled with what are called *limited end float couplings*. These couplings include a spacer that is installed between the two coupling halves that limit the end float of the motor to ensure that the motor rotor does not contact the bearing shoulders. The end float on 250- to 500-horsepower (HP) motors is limited to $1/4$ inch by National Electrical Manufacturers Association (NEMA) standards, and a $3/32$ limited end float coupling is used. On 500-HP and above motors, the end float is $1/2$ inch and a $3/16$ limited end float coupling is used.

Off-Zero Alignment

Just as a pump or turbine shaft gets longer when it heats up, the entire pump or turbine will expand. Alignment is, of course, intended to eliminate stresses when the unit is at operating temperature. Perfect alignment at operating temperature is frequently not the same as perfect alignment.

Drives for equipment that will run at high temperatures are frequently aligned with the drive centerline above the centerline of the driven machine. As the driven machine heats up, its centerline will come into line with the centerline of the drive. The amount of correction can be measured or approximated with the expansion formula, using the height of the shaft above the base in place of the shaft length.

In a few situations, a drive may have to be cold-aligned to correct angular misalignment. For example, a turbine coupling may be set up with an uneven gap if the bearing pedestals expand at different rates.

The first source of determining cold-alignment settings is from the manufacturer's literature. If cold-alignment settings are necessary, they are usually given in the maintenance and operating instructions. If cold settings are not provided, they can be approximated through the following calculation:

$$\text{Cold Offset} = H \times \frac{(T_P - T_m)}{1000} \times 0.0067$$

where

H = height of motor shaft centerline from feet

T_P = temperature of driven machine

T_M = temperature of motor

The correction factor of 0.0067 inch per 1000 inches is for most cast steels.

Consider the following situation. A pump operating at 250°F is driven by a 324T frame motor operating at 100°F. The offset can now be calculated using the following formula:

$$\text{Cold Offset} = 8'' \times \frac{(250 - 100)}{1000} \times .0067 = .008''$$

If vibration levels still indicate a misalignment condition after cold settings have been calculated, a more accurate graphical hot-alignment check is necessary.

In this method, indicator readings are obtained on the unit as soon as possible after the machine has been shut down. Several readings are taken, and the elapsed time since the machine was first shut down is noted for each set of readings. The following example explains this method.

Hot-alignment readings taken from motor to pump are entered into a table (Table C-1).

Table C-1   Hot-Alignment Results

Time	180 Degrees Reading ($^1/_2$TIR)
10:00 a.m.	unit shut down
10:15 a.m.	−.006 inch
10:30 a.m.	−.002 inch
10:45 a.m.	+.001 inch

These readings are plotted on graph paper as shown in Figure C-6.

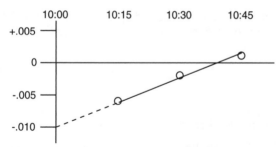

Figure C-6   Plotted hot-alignment results.

Drawing a line through the plotted points and extending the line to the time that the unit was initially shut down (10:00 a.m. in this example) projects where the machine was at that time. In this example, the motor was .010 inch low at operating conditions. The cold-alignment settings will have to be changed to adjust the motor .010 inch to ensure alignment at the operating temperatures.

Gear Drive Inspection Standards

Gear drive systems are best known for their ability to transmit high torque. Whereas belt (and even chain) drive arrangements are ulti-mately limited in the amount of torque they can transmit, gears are virtually unlimited in the amount of torque that they can transmit.

In practical terms, the loading of a gear set is a function of the following:

- The gear material
- The gear tooth profile, which in turn is a function of the gear pitch and pressure angle
- The working length of the gear face

Gear drives are also known for their long life. Properly lubricated and running at design load, gear sets have been known to operate satisfactorily after more than 100 years of operation.

Gear Tooth Nomenclature

Figure C-7 shows the important dimensions of a gear tooth and the nomenclature that identifies those dimensions. Following are some important terms:

- *Addendum*—The addendum is the radial height of a tooth above the pitch circle.

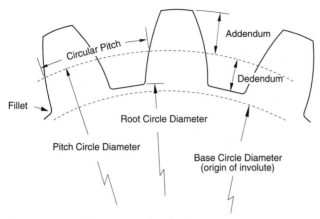

Figure C-7   Gear tooth nomenclature.

- *Clearance*—The clearance of a gear is defined as the radial distance from the top of a tooth to the root or bottom of the mating tooth space.
- *Dedendum*—The dedendum is the radial distance from the pitch circle to the root or bottom of a tooth. The dedendum is equal to the whole depth minus the addendum.
- *Fillet*—The fillet is a rounding of the base of the gear tooth at the root. This minimizes stress at that location.
- *Pitch*—The pitch of a gear is the distance, measured on the pitch circle, from one tooth to the same point on an adjacent tooth.
- *Outside Diameter*—The outside diameter of a gear is the overall diameter and is further specified as the pitch diameter plus two addendums.
- *Whole Depth*—The whole depth of a tooth is the radial distance from the top of a tooth to the root or bottom. It is equal to the addendum plus the dedendum.
- *Working Depth*—The working depth of a gear is the distance that a tooth extends into the tooth space of the mating gear. It is equal to two addendums.

Gear Maintenance

Someone has observed that "gears wear out until they wear in . . . and then they never wear out." The American Gear Manufacturers

Association (AGMA) describes this reliability of gear drive systems as follows:

> *It is the usual experience with a set of gears in a gear unit ... assuming proper design, manufacture, application, installation, and operation ... that there will be an initial 'running-in' period during which, if the gears are properly lubricated and not overloaded, the combined action of rolling and sliding of the teeth may smooth out the manufactured surface and give the working surface a high polish. Under continued proper conditions of operation, the gear teeth will then show little or no sign of wear.*

Although such an endorsement of reliability might make it seem that gear systems are indestructible or require no maintenance, this is definitely not the truth.

Gear Drive Backlash

Some provision for the clearance of mating gear teeth must be made for their proper operation. This provision is called *backlash*. Backlash is defined as the amount by which a tooth space exceeds the thickness of an engaging tooth. The backlash clearance compensates for machining tolerances, thermal growth of the gears at operating temperature, and any other factors that may interfere with the meshing of the gears.

Backlash of spur and helical gears is determined during manufacturing. The center distances of such parallel shaft arrangements is usually not adjustable. Backlash is normally measured by holding one gear motionless (usually the pinion) and rotating the other gear through the backlash clearance. With a dial indicator mounted against the face of any tooth, and the gear contacting one side of the mating gear, the indicator is zeroed. The gear is then moved in the other direction and the backlash is observed on the indicator.

Backlash can also be measured by inserting feeler gages between the gap between meshing teeth until the maximum gage is determined. Table C-2 and Table C-3 provide the recommended backlash for course-pitch and fine-pitch parallel shaft gearing.

The backlash for bevel gear arrangements must be set. This is accomplished by shimming one of the gears (or shafts) until the proper clearance is measured. Setting the backlash clearance on right-angle gear sets is usually not encountered unless a gear change has been made. Table C-4 gives the recommended backlash clearances for several right-angle drive gear sets.

Table C-2 AGMA Recommended Backlash Range for Coarse-Pitch Spur, Helical, and Herringbone Gearing

	Normal Diametral Pitches				
	0.5–1.99	2–3.49	3.5–5.99	6–9.99	10–19.99
Center Distance (Inches)	Backlash, Normal Plane (Inches)				
Up to 5					.005–.015
5 up to 10				.010–.020	.010–.020
10 up to 20			.020–.030	.015–.025	.010–.020
20 up to 30		.030–.040	.025–.030	.020–.030	
30 up to 40	.040–.060	.035–.045	.030–.040	.025–.035	
40 up to 50	.050–.070	.040–.055	.035–.050	.030–.040	
50 up to 80	.060–.080	.045–.065	.040–.060		
80 up to 100	.070–.095	.050–.080			
100 to 120	.080–.110				

Gear Failure Analysis

The gear industry defines two kinds of gear failure. One involves actual breakage of a tooth or teeth. The other involves surface deterioration of all the teeth.

Tooth Breakage

A missing or partially missing tooth (or teeth) evidences a tooth breakage failure. Sometimes, the cause of the breakage is easily identified by the presence of foreign matter that interfered with the gear mesh. Also, the surface of the break is lined or striated. When the cause of breakage is not readily evident, the surface of the fracture must often be closely inspected to determine the cause.

A broken tooth may have been a long time in developing. When a tooth has broken because of fatigue, the surface of the break is usually quite smooth. Following is some other evidence to look for when breakage has been the result of metal fatigue:

- *Point of origin*—In a fatigue failure break, there may be evidence of a beginning point to the break itself. Often, an inclusion or heat-treatment crack is the point or origin.

Table C-3 AGMA Backlash Allowance and Tolerance for Fine-Pitch Spur, Helical, and Herringbone Gearing

	Tooth Thinning to Obtain Backlash			
Backlash Designation	Normal Diametral Pitch Range	Allowance per Gear (Inch)	Tolerance per Gear (Inch)	Resulting Approximate Backlash, Normal Plane (Inches)
A	20 thru 45	0.002	0 to .002	.004 to .008
	46 thru 70	0.0015	0 to .002	.003 to .007
	71 thru 90	0.001	0 to .00175	.002 to .0055
	91 thru 200	0.00075	0 to .00075	.0015 to .003
B	20 thru 60	0.001	0 to .001	.002 to .004
	61 thru 120	0.00075	0 to .00075	.0015 to .003
	121 thru 200	0.0005	0 to .0005	.001 to .002
C	20 thru 60	0.0005	0 to .0005	.001 to .002
	61 thru 120	0.00035	0 to .0004	.0007 to .0015
	121 thru 200	0.0002	0 to .0003	.0004 to .001
D	20 thru 60	0.00025	0 to .00025	.0005 to .001
	61 thru 120	0.0002	0 to .0002	.0004 to .0008
	121 thru 200	0.0001	0 to .0001	.0002 to .0004
E	20 thru 60	ZERO	0 to .00025	0 to .0005
	60 thru 120		0 to .0002	0 to .0004
	121 thru 200		0 to .0001	0 to .0002

- *Fretting corrosion*—As a crack develops in a tooth, lubricating oil may seep into the crack. As the tooth meshes with the companion gear teeth, the oil is compressed and causes movement of the tooth metal. This, in turn, will set up a fretting corrosion condition that can be seen in the reddish or brown stain to a portion of the break surface metal.

- *Break location*—Tooth breakage caused by fatigue usually begins at or near the root of the tooth. This is because the stress on the tooth is greatest at this location.

Surface Deterioration
Surface deterioration of the gear teeth is the result of fatigue or lubrication failure. Where fatigue is the failure mode, several variations

Table C-4 AGMA Recommended Backlash Range
for Bevel and Hypoid Gears*

| Diametral Pitch | Normal Backlash, Inches | |
	Quality Numbers 7 through 13	Quality Numbers 4 through 6
1.00 to 1.25	0.020–0.030	0.045–0.065
1.25 to 1.50	0.018–0.026	0.035–0.055
1.50 to 1.75	0.016–0.022	0.025–0.045
1.75 to 2.00	0.014–0.018	0.020–0.040
2.00 to 2.50	0.012–0.016	0.020–0.030
2.50 to 3.00	0.010–0.013	0.015–0.025
3.00 to 3.50	0.008–0.011	0.012–0.022
3.50 to 4.00	0.007–0.009	0.010–0.020
4.00 to 5.00	0.006–0.008	0.008–0.016
5.00 to 6.00	0.005–0.007	0.006–0.013
6.00 to 8.00	0.004–0.006	0.005–0.010
8.00 to 10.00	0.003–0.005	0.004–0.008
10.00 to 16.00	0.002–0.004	0.003–0.005
16.00 to 20.00	0.001–0.003	0.002–0.004
20 to 50	0.000–0.002	0.000–0.002
50 to 80	0.000–0.001	0.000–0.001
80 and finer	0.000–0.0007	0.000–0.0007

*Measured at tightest point of mesh.

may be observed:

- *Pitting*—Pitting is the result of work hardening that occurs at the pitch line of gears. It usually appears on the pinion first. This is because the teeth of the pinion are seeing many more cycles of operation. Also, the pinion is generally the driver gear. When mating with a companion gear tooth, the compression of surface metal on the pinion is from the pitch line out to the edge of the tooth. This causes the surface stresses to be higher than if the compression were toward the root. Pitting is first observed as a line or area of roughness on the tooth surface (Figure C-8).

- *Vibration*—Vibration in gears causes momentary and cyclic high loading and resultant stresses on the gears. The vibration may be caused by the gears themselves or may be induced from

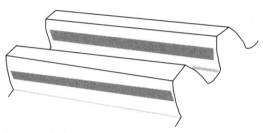

Figure C-8 Gear pitting.

driver or driven machinery. Vibration caused by the gears is usually a manufacturing problem. Unbalance, though rare in machined gears, can be the result of an inclusion in the initial forging. Vibration is usually caused by machining problems during the manufacturing process. If the gear blank was not perfectly placed on the gear cutting device, deviations in the gears cause the gear to travel axially back and forth with each rotation. In spur gears this is usually a minor problem, but it can be a major source of early fatiguing in helical gears. Induced gear vibration is usually the cause of misalignment of the gear to its adjacent machines. Also, failure of the coupling between the gear box and adjacent machinery can have the same effect.

- *Scoring*—Lubrication deficiencies usually cause gear surface failures commonly referred to as *scoring*. Tears or scratches appear on the tooth surfaces. They are perpendicular to the axis and usually cover the entire working surface of the gear teeth. Some scoring occurs during the break-in period of new gears. All gear boxes should be inspected internally after the first 100 hours of operation. Scoring caused during the break-in period usually stops after the surface imperfections are worn smooth. Frequent lubrication oil analysis after start-up can also detect break-in scoring and also verify if and when it begins to diminish (Figure C-9). The scoring caused during break-in can be minimized by requiring smoother tooth surface finishes during manufacturing. This comes at an increased initial cost, however, and should be weighed against the probability of this scoring stopping after a while.

V-Belt Drives

A belt drive transfers power from a driver (usually an electric motor) to another piece of equipment called the *driven machine*. Both the

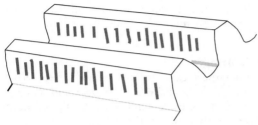

Figure C-9 Gear scoring.

driver and the driven equipment are equipped with a grooved wheel commonly called a sheave and sometimes called a pulley. The belt connects the sheaves and causes the driven sheave to turn with the driver (Figure C-10).

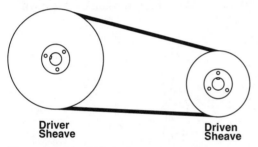

Figure C-10 V-Belt drive arrangement.

Causes of Belt Slip

Belt slip is movement of the belt against the sheave. Some amount of slip is unavoidable. *Belt creep*, which involves a slight forward movement of the belt, cannot be eliminated. However, on a well-designed and properly tensioned drive, slip is generally less than 1 percent. If slip is excessive, a loud squealing sound will be heard and belt wear will be accelerated. The following are causes of excessive belt slip:

- Incorrect tension
- Sheave wear and damage
- Alignment
- Belt tensioning

Incorrect Tension

This is the most common cause of belt slip. Tension applied when the drive is stopped must be sufficient to transmit the power required of the drive when it is running. A tightening method using a belt tension gage or the elongation method should be used to ensure proper tension.

Sheave Wear and Damage

Burrs or rough spots along the sheave rim can damage a belt. When changing a belt, sheaves should be inspected for wear and damage. They should be cleaned thoroughly to remove dust, oil, and other foreign matter that can lead to pitting and rust.

Use a sheave gauge (available from a belt manufacturer) to check the sides of the sheave groove for wear. Sheave walls should not be dished out or worn. A shiny groove bottom suggests either that the sheave is so worn that the belt is bottoming out in the groove or that the wrong size or design belt is being used.

All installed sheaves should be checked for eccentricity (outside diameter run-out) or wobble (face run-out), as shown in Figure C-11.

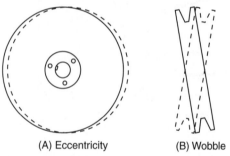

(A) Eccentricity   (B) Wobble

Figure C-11   Sheave eccentricity and wobble.

An improperly bored hub, bent shaft, or improperly installed bushings are the common causes for these conditions. Table C-5 describes the standards that should be used when checking for wobble or eccentricity.

A dial indicator mounted on a magnetic base is often used for this inspection.

Alignment

There are basically three forms of misalignment, as shown in Figure C-12.

Alignment can be checked with a length of string. Tie off or hold the string across the length of the two sheaves near the center of the sheaves. If two shafts are parallel, the string will touch both edges

Table C-5   Acceptable Standards for Eccentricity and Wobble

	Sheave Diameter	*Max Acceptable Run-Out*
Radial Run-Out (eccentricity)	Up through 10 inches	.010 inch
Limits for Sheaves:	For each additional inch of diameter	Add .0005 inch
Axial Run-Out (wobble):	Up through 5 inches	.005 inch
	For each additional inch of diameter	Add .001 inch

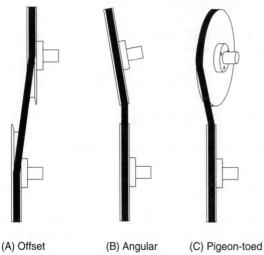

(A) Offset   (B) Angular   (C) Pigeon-toed

Figure C-12   Three forms of sheave misalignment.

on both sheaves (four points) at the same time. Move the shafts into parallel alignment if necessary.

Hold the string nearer the top edge of the sheaves and attempt to touch the four edges. If the string does not touch four points, shim up the front or back of the driver (usually a motor) until it does.

Belt Tensioning

The most frequent cause of belt drive failure is loose belts. Slipping belts heat up excessively and will eventually break.

Ideal belt tension is just at the point where the belts stop slipping. This is the point where the wedging action force is equal to the friction force holding the belt on the sheave. Attaining this point, however, is impractical because the belt will stretch slightly during normal operation. Attempts to properly tension a belt drive for operating conditions may result in excessive tension, causing high forces on the bearings and possibly breaking the belts. Determination of the optimum tension is best attained using the following force deflection method:

1. Evenly pretension the belts, attempting not to disrupt the drive alignment, until they are snug.

2. Measure the distance from the point where the belt first touches one sheave to the same point on the other. (This length is approximately equal to the distance between the shaft center lines.)

3. Determine the test deflection by multiplying the above length by $1/64$ inch (.015625 inch).

4. Using the spring scale, deflect the center point on the belt the distance calculated previously and record the force required in pounds.

5. Measure the pitch diameter (P.D.) of the small sheave or find the P.D. engraved on the sheave.

6. Determine the speed of the small sheave (RPM).

7. Determine the speed ratio.

8. Use the manufacturer's charts to determine the correct tension range.

9. Tighten or loosen the belts and recheck the force (Step 4) until the force required is within the range provided in the chart.

10. When installing new belts it is good practice to tension them with a deflection force $1/3$ greater than the maximum force recommended.

11. Recheck the tension again at the end of the first day of operation.

12. For banded belts multiply the forces recommended for the number of belts and tighten to that force.

Pumps and Mechanical Seals

When equipment is out of service for a shutdown, you should pay particular attention to pumps and mechanical seals. This includes the following areas worth noting:

• System curves
• Pump characteristic curves

- Combined system and pump curves
- Pump affinity laws
- Cavitation
- Pump condition
- Pump repair
- Mechanical seals

System Curves

Every liquid-transfer system has a unique head-capacity (pressure-flow rate) curve that represents the amount of energy that must be put into a system to move liquid. This curve is called the *system curve*.

The *liquid-transfer system* is defined as the point where the liquid is being pumped from, the piping and valves up to the pump, and the piping and valves from the pump discharge, to the point where the liquid is discharged. To completely define the system, the elevation difference between both the starting point and ending point must be known, as well as the relationship between both points and the pump itself. Finally, the pressure at both the starting and end point must be known.

Static Head

Static head is the minimum head the pump must develop just to begin to move liquid. If the starting point and ending point are both under atmospheric pressure, the static head is simply the elevation difference between the two points, as shown in Figure C-13.

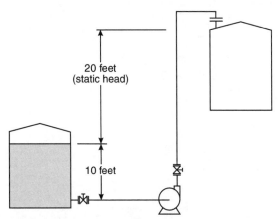

Figure C-13   Static head.

The starting point is 10 feet above the pump suction. If all the valves are open in the discharge piping, then the liquid level in the piping is 10 feet above the pump. The ending point is another tank, an additional 20 feet above this point. The pump must develop at least 20 feet of head just to get liquid up to the ending point. If the pump only develops 20 feet of liquid head, no liquid will be moved. The pump will be able to lift water up to the higher tank but will be unable to pump into the tank.

If the starting point had been 10 feet lower than the pump, the pump would have to develop a minimum of 40 feet of head just to get liquid up to the ending point. Additionally, if the ending point had been under pressure, the pump would also have to overcome that pressure in addition to the vertical lift.

In short, the static head determines the minimum head the pump must develop before it will begin to move liquid. The static head identifies the point on the system curve where the flow rate is zero.

Piping Losses

The remainder of the system curve is determined by the *piping losses*. These are the friction losses that will occur throughout the system as liquid flows. The losses increase as the flow rate of liquid is increased. As mentioned before, these losses are determined by friction coefficients multiplied by the velocity head. The losses increase by the square of the flow rate. At no flow, there are no friction losses. Each time the flow is doubled, the friction losses increase four times. Figure C-14 shows a typical system curve.

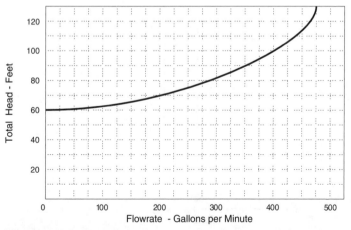

Figure C-14 System curve.

Friction losses in a piping system can change. Clean piping will have lower losses than extremely fouled piping. On complex pumping systems, several system curves may be determined to correspond to several fouling levels. The hydraulic engineer can then evaluate several pumps. The engineer may determine it to be more economical to clean the piping more often and use a less-expensive pump.

The system curve for a specific system is independent of the liquid being pumped. Even though the curve may be determined using water as the pumped fluid, the curve would hold true for any other liquid. This is the major reason that the curve is determined as a function of the liquid head as opposed to other pressure units.

Pump Characteristic Curves

Each centrifugal pump, with a given impeller size and operating at a given speed, will have a characteristic curve that will define how much total head (in feet of liquid) it will develop at various flow rates (usually in gallons per minute). As with the system curve, the *pump curve* will hold true regardless of the liquid pumped. Figure C-15 shows a typical pump characteristic curve.

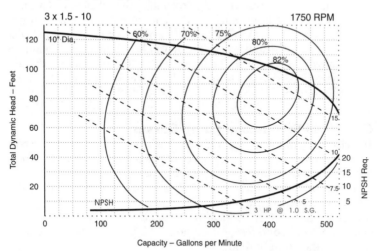

Figure C-15   Pump curve.

The important portions of the curve to recognize are:

- *Total Dynamic Head (TDH)*—This is the total liquid head the pump will develop. At no flow, it determines the pump's maximum liquid head. This point on the pump curve must at least exceed slightly the static head of the system curve or no liquid will be pumped.

- *Net Positive Suction Head (NPSH)*—This is the minimum positive head (measured in absolute value) the pump must have at the eye of the impeller. Otherwise, the pump will cavitate. Cavitation will be discussed in more detail later.

- *Brake Horsepower (BHP)*—This is the horsepower that must be delivered to the pump if it were pumping a liquid with a density equal to water. If the density of the liquid being considered is different than water, the BHP value from the curve is multiplied by the ratio of the liquid's density to the density of water. For example, the density of sulfuric acid is 1.86 times that of water. If the pump curve under consideration indicated that 10 BHP were needed to pump water, 18.6 BHP (1.86 × 10 HP = 18.6 HP) would be needed to pump the acid. The next largest motor of 20 HP would be chosen, as reducing the motor size will overload the motor. It should be noted that the BHP curve takes into account the effect of internal losses and the efficiency of the pump itself.

- *Efficiency*—Most pump curves will also indicate the pump's efficiency. Expressed as a percentage, the efficiency is a measurement of how much of the energy put into the motor is converted to pumping energy. In evaluating the operating cost of a pump installation, the hydraulic engineer will want to know what the overall efficiency is. A different pump, though more expensive, may have a higher efficiency and be more economical.

Combined System and Pump Curves

To select a pump, the hydraulic engineer superimposes the system curve onto the characteristic curves of the pumps being considered. Figure C-16 shows the two curves together.

The point where the two curves cross each other is called the *operating point*. For the system and pump curves shown, the pump will develop 100 feet of head at a capacity of 410 GPM with a minimum NPSH of 8 feet. A standard 20 HP motor should be used to supply the 16 BHP required.

In practical applications, the engineer will pick a pump that will provide an operating point well above the required flow rate and then plan to reduce the flow with a throttling valve. If fouling of the system occurs (thus increasing the line losses), the throttling valve can be opened to maintain the desired flow.

The suction requirements of pumps are often overlooked. On any pump being considered, the net positive suction head available

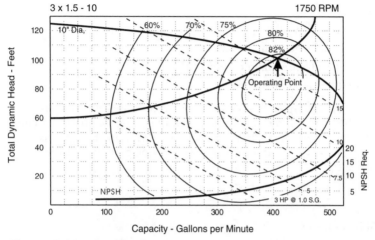

Figure C-16   Combined system and pump curve.

(NPSHA) must be calculated. This is simply the beginning point absolute head pressure plus the suction side liquid height minus the line losses in the suction piping. The available head must exceed the net positive suction head required (NPSHR) or else the pump will cavitate.

Pump Affinity Laws

Even with the many sizing options provided by manufacturers, an ideal operating point may not be found on existing pump curves. At this point, the engineer must entertain the options of modifying an impeller size or operating the pump at a specific speed different than common motor speeds. The pump affinity laws can be used to determine what is the best option.

Within the same pump casing, the following relationships apply to a change in impeller size (assuming no change in efficiency):

$$\frac{Q_2}{Q_1} = \frac{D_2}{D_1}$$

$$\frac{H_2}{H_1} = \left(\frac{D_2}{D_1}\right)^2$$

$$\frac{BHP_2}{BHP_1} = \left(\frac{D_2}{D_1}\right)^3$$

where

Q = Flow rate

D = Impeller diameter

H = Total dynamic head

BHP = Brake horsepower

As an example, assume that a 10-inch impeller in a Durco $3 \times \frac{1}{2}-10$ is too large. At the current operating point, it develops 100 feet of head, its flow rate is 150 GPM, and the BHP requirement is 15 HP. What would be the operating point with an $8\frac{1}{2}$-inch impeller?

$$\frac{Q_2}{150\,\text{GPM}} = \frac{8.5''}{10''}$$

Solving for Q_2,

$$Q_2 = 150\,\text{GPM} \times \frac{8.5''}{10''} = 127.5\,\text{GPM}$$

$$\frac{H_2}{100\,\text{ft}} = \left(\frac{8.5''}{10''}\right)^2$$

Solving for H_2,

$$H_2 = 100\,\text{ft} \times \left(\frac{8.5''}{10''}\right)^2 = 72.3\,\text{ft}$$

$$\frac{BHP_2}{15\,\text{HP}} = \left(\frac{8.5''}{10''}\right)^3$$

Solving for BHP_2,

$$BHP_2 = 15\,\text{HP} \times \left(\frac{8.5''}{10''}\right)^3 = 7.8\,\text{HP, use a 10 HP motor.}$$

With a smaller diameter impeller, less flow and pressure will be developed, but a smaller motor can be used if this service is acceptable.

The affinity laws can also be used for a speed change. These relationships apply (assuming no change in efficiency):

$$\frac{Q_2}{Q_1} = \frac{N_2}{N_1}$$

$$\frac{H_2}{H_1} = \left(\frac{N_2}{N_1}\right)^2$$

$$\frac{BHP_2}{BHP_1} = \left(\frac{N_2}{N_1}\right)^3$$

where,

$$N = \text{speed of pump}$$

In addition,

$$\frac{NPSHR_2}{NPSHR_1} = \left(\frac{N_2}{N_1}\right)^3$$

where,

$$NPSHR = \text{Net positive suction head required}$$

As an example, assume that a 10-inch impeller in a Durco $3\times \frac{1}{2}-10$ is too small. At the current operating point it develops 100 feet of head, its flow rate is 150 GPM, and the BHP requirement is 15 HP. It was decided to speed the pump up from 1750 RPM to 2250 RPM. The NPSH requirement is currently 8 ft. What would be the pump's new characteristics? Is the existing motor sufficient for the job?

$$\frac{Q_2}{150\,\text{GPM}} = \frac{2250}{1750}$$

Solving for Q_2,

$$Q_2 = 150\,\text{GPM} \times \frac{2250}{1750} = 193\,\text{GPM}$$

$$\frac{H_2}{100\,\text{ft}} = \left(\frac{2250}{1750}\right)^2$$

Solving for H_2,

$$H_2 = 100\,\text{ft} \times \left(\frac{2250}{1750}\right)^2 = 165\,\text{ft}$$

$$\frac{BHP_2}{15\,\text{HP}} = \left(\frac{2250}{1750}\right)^3$$

Solving for BHP_2,

$$BHP_2 = 15 \text{ HP} \times \left(\frac{2250}{1750}\right)^3 = 32 \text{ HP} \text{ Use a 40 HP motor.}$$

$$\frac{NPSHR_2}{8 \text{ ft}} = \left(\frac{2250}{1750}\right)^3$$

Solving for $NPSHR_2$,

$$NPSHR_2 = 8 \text{ ft} \times \left(\frac{2250}{1750}\right)^3 = 17 \text{ ft}$$

Operating at the higher speed will provide a higher flow and pressure but will require a much larger motor and more NPSH.

Understanding Cavitation

The pressure of the liquid determines the boiling temperature. Water at sea level is under an atmospheric pressure of 14.7 psi and will boil at 212°F. If the atmospheric pressure is reduced, the boiling point temperature is lower. If the pressure is low enough, water can be made to boil at room temperature.

Cavitation is simply the result of the pumped liquid boiling at the eye of the impeller. Boiling occurs because the absolute pressure at the eye is very low. The vapor bubbles developed from the boiling liquid implode at the higher pressures away from the eye. This causes the characteristic popping noise often heard in a cavitating pump.

An examination of Bernoulli's equation tells us that the pressure at the eye of the impeller is determined by the suction pressure:

$$\text{Suction Pressure} = P_s + H_s - \left(K_s \times \frac{v_s^2}{2g}\right)$$

In short, cavitation occurs when the friction losses in the suction piping (Ks times the velocity head) are large enough to offset the positive pressure (Ps) at the suction tank plus any liquid height (Hs) in the tank relative to the pump suction flange. Bernoulli's equation, however, also gives insight as to the possible corrections to eliminate cavitation:

- *Raise the level in the tank*—This would increase the liquid head pressure (more Hs) relative to the pump.

- *Increase the pressure in the suction tank*—This is not often an option, but it would increase pressure (more Ps) relative to the pump.

- *Increase the suction pipe size and eliminate unnecessary suction piping restrictions (less Ks)*—This is often the correction of choice. Minimizing the friction losses that are offsetting the positive pressures raises the pressure at the pump suction. Doubling the suction pipe diameter reduces friction losses by four times as the velocity head is halved.

Determining Pump Condition

Bernoulli's equation also provides a means to evaluate centrifugal pump condition. This evaluation is derived by considering the equation when the pump is operating at dead headed or no flow conditions (pump discharge valve is closed). Because there is no flow, both velocity head terms are zero and the equation then becomes:

$$E\,(TDH) = (Pd - Ps) + (Hd - Hs)$$

Since the liquid heights are measured relative to the pump, the discharge liquid height is essentially zero. To make sure that all terms have equal units, the pump discharge and suction pressures must be converted from psi units to the equivalent feet of liquid height:

$$E\,(TDH) = Pd - Ps - Hs$$

The calculated value of TDH can be compared to the zero flow point on the pump characteristic curve. If it is less than what the pump requires (as indicated on the curve), impeller or volute wear may be evident. The pump will have to be overhauled to restore capacity.

Centrifugal Pump Repair

The following overhaul methods are general to centrifugal pumps. Some procedures may vary slightly from one specific manufacturer to another. Always consult the correct maintenance and repair manual for the make and model that you are overhauling.

Pump overhaul should always be performed in a clean, well-lit shop. Don't attempt to perform overhaul procedures in the field.

Disassembly

Many pump manufacturers offer a pump design known as a standard American National Standards Institute (ANSI) back pull-out design. The volute casing is designed so that it can remain installed to the piping flanges, whereas the power end of the pump can be removed for service. If a spacer drop-out type of coupling is used as well, the drive motor does not have to be disconnected and moved either.

If the pump conforms to this standard, the following procedure will prepare the pump unit for shop overhaul:

1. Remove spacer drop-out coupling. If the pump is not coupled with a spacer drop-out type coupling, it will be necessary to uncouple and move the motor away from the pump.

2. Remove the casing nuts that attach the back plate to the volute casing.

3. At the bearing housing, disconnect the rear foot mount from the housing. If possible, leave the rear foot tightened down to the base plate, as this will help support the pump assembly as it is moved away from the volute casing.

4. When pump assembly is clear of the volute, remove the assembly from the base. Take care not to damage the impeller or the back plate mating flange when setting the assembly down. Remove the rear foot mount from the base plate and reattach to the bearing housing.

5. Move the pump assembly to the shop for inspection and overhaul.

6. When the overhaul is complete, or when a rebuilt unit is being installed in its place, these procedures are followed in reverse order.

Shop Measurements

Before the pump is disassembled, a number of measurements should be taken. These measurements will provide insight as to the possible problems that will require correction during the overhaul process.

The assembly should be bolted or clamped to a sturdy workbench. The rear foot mount will come in handy here. The following measurements should be taken and recorded, either on a pump repair data sheet or on the work order itself.

Shaft End Play Excessive end play will cause premature seal failure and will also limit the overall life of the pump bearings, especially the thrust bearing. Excessive end play before disassembly is usually caused by damage to the thrust bearing or the components that hold it within its housing. To check for shaft end play, use the following procedure:

1. Using a soft hammer, tap the shaft toward the rear of the pump.

2. Install a dial indicator on the pump with the dial stem contacting the shaft shoulder (Figure C-17).

3. Zero the dial indicator.

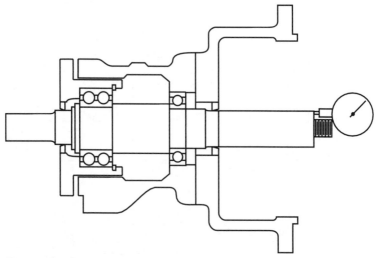

Figure C-17   Pump shaft end play.

4. Now move the shaft toward the indicator, tapping gently on the coupling end of the shaft.

5. Record the indicator reading. End play should not exceed .004 inch.

Bent Shaft A bent shaft will also limit seal and bearing life. A bent shaft indicates severe operating conditions, or could have been the result of improper assembly during a prior repair. To check for a bent shaft, use the following procedure:

1. Mount a dial indicator on the pump housing with the dial stem reading on the shaft (Figure C-18).

2. Zero the dial and rotate the shaft through one revolution. Record the total needle travel (total indicator run-out, or TIR).

3. Total indicator run-out should not exceed .002 inch. The shaft should be checked at several locations on the wetted end.

4. Check the coupling end for run-out as well. Take care not to damage the indicator on the shaft key-way.

Stuffing Box Squareness If the stuffing box face is not square, the mechanical seal gland will seat unevenly when installed. Under operation, the rotating assembly of the seal will undergo excessive axial movement as it attempts to seal against the tilted stationary face.

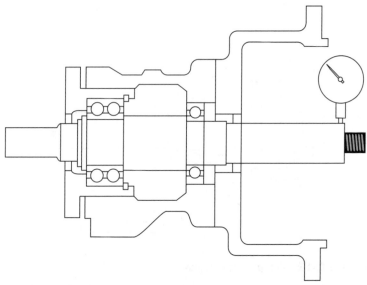

Figure C-18 Bent pump shaft check.

Obtaining this measurement requires a special indicator that is small enough and designed to obtain such readings. It is commonly referred to as a *last word indicator*. To check for stuffing box square-ness, use the following procedure:

1. Attach the stuffing box cover to the back plate.
2. Clamp a dial indicator to the pump shaft, with the indicator stem squarely setting on the box cover (Figure C-19).
3. Zero the indicator and rotate the shaft through one revolution. Record the total indicator run-out. It should not exceed .002 inch or .001 inch per inch of shaft thickness, whichever tolerance is greater.

Stuffing Box Concentricity This is a difficult measurement to take, but again it must be checked on pumps using a mechanical seal. (A last-word indicator is indispensable here.) Use the following procedure:

1. Mount a dial indicator on the pump shaft, with the indicator stem sweeping the inside bore of the stuffing box. A wiggler may make this easier (Figure C-20).
2. Zero the indicator and rotate the shaft one revolution. Record the total indicator run-out. It should not exceed .005 inch.

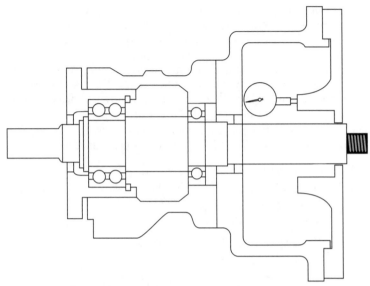

Figure C-19   Stuffing box squareness check.

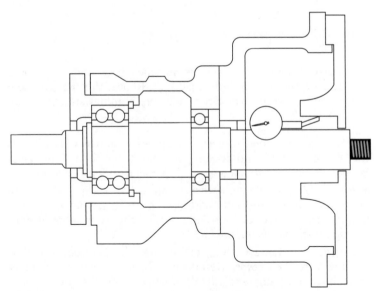

Figure C-20   Stuffing box concentricity check.

Once the foregoing measurements have been taken, the pump can be disassembled. Before disassembly can begin, the bearing housing will have to be drained of oil. Carefully inspect the oil, examining a sample in a glass jar or beaker. The oil should not be cloudy, indicating moisture contamination. The oil should not have any objectionable odor, indicating possible chemical contamination.

When the pump is completely apart, the following measurements should be taken on the components.

Shaft Straightness If the shaft run-out was excessive, the problem could be a bent shaft. To check, follow this procedure:

1. Install the shaft between steady rests.
2. Mount a dial indicator on the shop table with the indicator stem on the shaft (see Figure C-21).

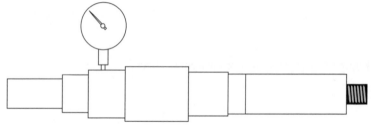

Figure C-21 Pump shaft straightness check.

3. Zero the indicator and rotate the shaft one revolution. Record the total indicator run-out. Repeat this measurement at several locations on the shaft. No readings should exceed .002 inch.

Shaft Seal Area Inspection Inspect the shaft location where the oil seals ride. There should be no discernable wear of the shaft. Damage at these areas can be corrected with spray metalizing procedures.

Shaft Seat Tolerance Both the thrust and radial bearings should be a light press fit to the shaft. With an outside micrometer, measure the shaft seat at two locations and at 90° from each other at both bearing locations (see Figure C-22).

Housing Bore Tolerance Both the thrust and radial bearings should be a sliding fit to their respective housing. With an inside micrometer (or a telescoping gauge and outside mike), measure the housing bore at two locations and at 90° from each other.

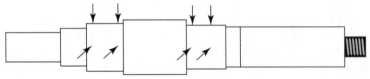

Figure C-22   Pump shaft seat tolerance.

Unless the pump maintenance and operating manuals indicate otherwise, Table C-6 should be used in evaluating bearing fits.

Table C-6   Bearing Shaft and Housing Fits

Bearing Bore (mm)	Shaft Seat to Bearing Tolerance** (Slide Fit) Fit in .0001 inch*	Bearing O.D. (mm)	Housing to Bearing Tolerance** (Light Force Fit) Fit in .0001 inch*
20	1T to 5T	32 to 47	6L to 0
25	2T to 6T	52 to 80	7L to 0
30 to 40	1T to 5T	85	10L to 0
45	2T to 6T	90 to 120	9L to 0
50	1T to 5T	125	11L to 0
55 to 60	1T to 6T	130 to 140	10L to 0
65	2T to 7T	145	11L to 0
70 to 80	1T to 6T	150 to 180	10L to 0
85	2T to 8T		
90 to 100	1T to 7T		

*T and L stand for tight and loose fit, respectively.
**Out-of-round or taper tolerance should not be more than $^1/_2$ the fit tolerance.

Reassembly
Reassembly is generally a reversal of the procedures used in disassembly. Again, it is imperative that the correct shop manual be used for the make and model you are repairing.

Impeller Adjustments
One adjustment critical to the pumps operation is the location of the impeller relative to the volute casing. This dimension ensures that the pump will operate at the rating of its characteristic curve.

Depending on the manufacturer, this dimension is generally determined by either the impeller's location relative to the front of the volute casing or the impeller's location relative to the casing back plate.

Figure C-23 shows a sample of a typical inspection sheet that should be filled out by a maintenance worker who performs pump inspections.

Centrifugal Pump
Repair Data Sheet

Equipment ID_____W/O No. _____Manufacturer_____Model_____S/N_____

As Found Alignment Data (Motor to Pump Readings)

Rim Readings: 0_____   90_____   180_____   270_____

Face Readings: 0_____   90_____   180_____   270_____

Installation Inspection (note any problems with base, groul, piping, motor or coupling):

_____

_____

Shop Measurements (prior to disassembly):
Shaft End Play _____   Shaft Run-out: Impeller End_____   Seal Area_____   Coupling End_____

Stuffing Box Squareness _____   Stuffing Box Concentricity _____

Shop Measurements (disassembled):
Shaft Run-out   Impeller End_____   Line Bearing_____   Thrust Bearing_____ Coupling End_____

Bearing Data: Thrust Bearing_____   Bore_____ Line Bearing_____   Bore _____

Shaft Seat Dimensions:
Thrust Bearing Seat: Shoulder End: 0-180_____ 90-270_____   Seal End: 0-180_____ 90-270_____
                     Shaft Seat Out of Roundness_____   Shaft Seat Taper_____

Line Bearing Seat: Shoulder End: 0-180 _____ 90-270 _____   Seal End: 0-180 _____ 90-270_____
                     Shaft Seat Out of Roundness_____   Shaft Seat Taper_____

Bearing Housing Dimensions:
Thrust Housing: Shoulder End: 0-180 _____ 90-270_____   Seal End: 0-180_____ 90-270_____
                    Housing Out of Roundness_____   Housing Taper_____

Line Housing: Shoulder End: 0-180 _____ 90-270_____   Seal End: 0-180_____ 90-270_____
                    Housing Out of Roundness_____   Housing Taper_____

Shop Measurements (after assembly):
Shaft End Play_____   Shaft Run-out: Impeller End_____   Seal Area_____   Coupling End_____

Stuffing Box Squareness_____   Stuffing Box Concentricity_____

Final Alignment Data (Motor to Pump Readings):
Rim Readings: 0_____   90_____   180_____   270_____

Face Readings: 0_____   90_____   180_____   270_____

Figure C-23   Centrifugal pump worksheet.

Mechanical Seals

Mechanical seals have all but replaced traditional packing and lip seals in difficult sealing applications. The largest usage still remains in centrifugal pump installations. In pumping applications, mechanical seals are preferred where pumped fluids are hazardous, toxic, or must be kept contaminant-free, such as foods and pharmaceuticals.

Mechanical seals have been successfully installed to operate in pressures up to 3000 psig, rotating speeds up to 50,000 RPM, and temperatures ranging from cryogenic up to 1200°F. When properly selected, installed, and operated, mechanical seals have been documented to last up to five years before replacement was necessary. The disadvantages of mechanical seals are their inability to handle axial movement and their high initial cost.

Although considered leak-free, mechanical seal design requires that the two mating seal faces be lubricated with an extremely thin layer of fluid. If this fluid is the pumped medium, the resulting environmental release of fluid (though measurable) is so minute that the amount is considered negligible.

Troubleshooting Mechanical Seal Problems

When a mechanical seal fails, leakage can only be caused by one of two reasons. Either the leakage is from the primary seals themselves, or the secondary seals have failed instead.

Many seal problems remain unsolved because the evidence needed to find the solution is thrown away when a seal fails. The best method to solving seal problems is to do a postmortem on the failed seal. Here's what to look for.

Inspection of the Seal Faces

The wearing face of a mechanical seal is the only expendable part of the seal. Inspection of this component first will tell if the seal has lasted a normal life or not. The extension of the wearing face of most primary seals is about .125 inch. If this extension is worn down to the seal ring body, the seal is worn out. If extensive wear material is still present, the seal failed prematurely. Inspection of the primary seal faces and the wear patterns observed can often identify the cause of early failure. Although the wider primary face does not actually wear, the contact pattern from the narrower face is evident on the face of the seal.

Following are some things to look for:

- *Normal wear pattern*—The wear pattern shown on the stationary face in Figure C-24 is what should be seen. The width of the wear pattern should be little more than the width of the narrow face.

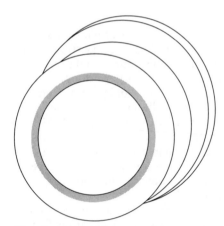

Figure C-24   Normal wear pattern.

- *Wide wear pattern*—The pattern in Figure C-25 indicates that the narrow face is sliding on the wide face. This is caused by excessive misalignment of the pump to the driver.

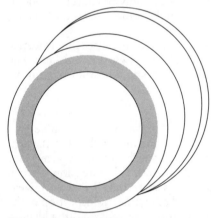

Figure C-25   Wide wear pattern.

- *Intermittent or uneven wear*—The wear pattern shown in Figure C-26 indicates that the gland plate is distorting the wide face. Check the gland face for run-out and trueness. Ensure that the gland nuts are tightened evenly.

- *Chipped edges*—The seal face in Figure C-27 is chipped, which means that the two faces have been hitting each other. The most

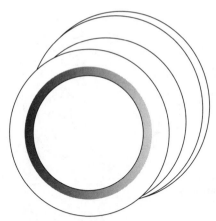

Figure C-26   Intermittent or uneven wear.

common cause of this condition (provided that the face wasn't damaged when it was installed) is cavitation within the pump.

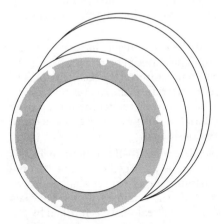

Figure C-27   Chipped edges.

- *Uneven wear patterns*—The uneven wear patterns shown in Figure C-28 are also caused by misalignment.

The biggest single cause of premature seal failure is misalignment between the pump and its driver. A rigid alignment standard and strict adherence to that standard is the single most important effort that can dramatically increase mechanical seal life.

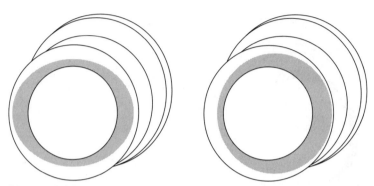

Figure C-28   Uneven wear patterns.

Inspection of the Elastomer

Although failure of the seal usually occurs at the primary seal (the faces), failure can also occur at the secondary seals. Inspection of the elastomeric members might indicate the following.

- *Swollen elastomer*—If the elastomer is swollen, soft, or sticky, it is incompatible with the material being pumped. Change to a different material.

- *Loss of compressibility*—If the elastomer is hardened, it too can be caused by attack from the material being pumped. It can also be caused by excessive temperature. The easiest cure is to again change to a different material. If the operating temperature of the seal environment exceeds the temperature rating of the secondary seal material, a flush or barrier system with external cooling might be considered.

- *Mechanical damage to secondary seals*—If the shaft surface is rough, the secondary seals can be damaged or can be become stuck to the shaft. Seal manufacturers expect shaft finish to be smooth enough to preclude such failures. Cartridge seals eliminate this type of failure as the seal sleeve is finished by the manufacturer.

Inspection of Other Seal Components

Some early seal failures are caused by problems with the seal body itself. Material buildup on the compression springs can cause them to hang up and not exert proper seal face loading. Such buildup is usually caused by excessive temperature within the stuffing box. Switching to a double seal arrangement and providing a cooled barrier flush can solve such failures.

Fans

The movement of gas is the second most common mechanical process in industry, next to liquid pumping. Industrial gas-moving equipment falls into three categories:

- *Compressors*—Used to develop high pressures, up to 500 psi. They are mostly used to raise the static pressure of a gas.

- *Blowers*—Considered for air moving requirements up to 10 psi. Blowers are used primarily to provide air for dry material transfer.

- *Fans*—Considered for low-pressure applications below 1 psi. One psi is equal to 30 inches of water (1 psi = 30 in of H_2O). Fans have a wide range of uses from ventilation to material transfer.

Fans are relatively simple to apply, although knowledge of design principles does provide additional insight that can prove valuable. This training session concentrates more on the operation and maintenance of a fan system rather than the design of a fan system. Special emphasis is placed on correcting problems caused by conditions of operation that were not considered by the system designer. In general, fan output is determined by the following:

- Speed (RPM)
- Diameter and width of the wheel
- Curve, pitch, and size of the blades
- Design of the housing

Fans are divided into two general classes: centrifugal and axial. The design of *centrifugal fans* causes air to move in a circular motion within the housing. The circular motion and the weight of the air create a centrifugal force, which is the pressure that moves the air. Air enters along the axis for most centrifugal fans. In all but the tubular type fan, the air flow changes to a radial direction.

Unlike most centrifugal fans, air flow is straight through an *axial fan*. Unlike centrifugal fans, which use the circular motion of the gas to create a pressure, axial fans add energy to the gas by pushing it through the fan. The housing tends to direct the flow along the axis, hence the name.

Fan Laws

When a fan manufacturer develops a fan, it publishes one characteristic curve for one speed, a standard gas density, and a standard temperature. The use of a series of equations (called the *fan laws*) can help a designer or user predict the performance of a fan under other conditions.

These equations are derived from the theories of fluid mechanics and include some assumptions and approximations. More exact data can be achieved using the classical formulas, but for most practical applications, the fan laws are sufficient.

One example of the use of the fan laws involves the increase or decrease of fan output from the published characteristics. The fan laws provide a method of determining how a change in speed can achieve the desired result. The following equations show the fan laws as they relate to a speed change.

With fan size and gas density remaining the same:

$$\frac{CFM_2}{CFM_1} = \frac{RPM_2}{RPM_1} \quad \frac{TP_2}{TP_1} = \left(\frac{RPM_2}{RPM_1}\right)^2 \quad \frac{SP_2}{SP_1} = \left(\frac{RPM_2}{RPM_1}\right)^2$$

$$\frac{BHP_2}{BHP_1} = \left(\frac{RPM_2}{RPM_1}\right)^3 \quad \frac{VP_2}{VP_1} = \left(\frac{RPM_2}{RPM_1}\right)^2$$

where

CFM is the flow rate in cubic feet per minute

SP is the specific pressure in pounds per square inch (PSI)

VP is the velocity pressure in pounds per square inch (PSI)

TP is the total pressure in pounds per square inch (PSI)

BHP is brake horsepower, which is the horsepower requirement of the fan

Subscript $1(_1)$ refers to design or inital conditions, Subscript $2(_2)$ designates the converted condition.

Most fan characteristic curves are provided for gas density of 0.075 lbs/ft^3, which is the density of dry air at sea level and at a temperature of 70°F. Standard tables, such as those found in *Perry's Chemical Engineers' Handbook* (New York: McGraw-Hill Professional, 1997) should be used if other gasses are being moved and for different gas temperatures.

Dry air at higher altitudes will also result in a lower gas density. In general, gases at higher temperatures have a lower density than the same gas at a lower temperature, and wet gases have a higher density than dry. If the density of gas to be moved is different than the one used to define the fan characteristics, the following fan laws apply. This is described by the following relationships:

Fan size and speed remaining constant:

$$CFM_1 = CFM_2 \qquad \frac{TP_2}{TP_1} = \frac{d_2}{d_1}$$

$$\frac{SP_2}{SP_1} = \frac{d_2}{d_1}$$

$$\frac{VP_2}{VP_1} = \frac{d_2}{d_1} \qquad \frac{BHP_2}{BHP_1} = \frac{d_2}{d_1}$$

d is the density of the gas.

Finally, although less usable for an installed fan, the effect of a change in fan size is shown in the following relationships:

Fan speed and gas density held constant:

$$\frac{CFM_2}{CFM_1} = \left(\frac{D_2}{D_1}\right)^3 \qquad \frac{TP_2}{TP_1} = \left(\frac{D_2}{D_1}\right)^2$$

$$\frac{SP_2}{SP_1} = \left(\frac{D_2}{D_1}\right)^2 \qquad \frac{BHP_2}{BHP_1} = \left(\frac{D_2}{D_1}\right)^5$$

$$\frac{VP_2}{VP_1} = \left(\frac{D_2}{D_1}\right)^2$$

D is the diameter of the fan.

Fan Performance Curves

Tests performed at the factory are used to develop fan characteristic curves. The Air Movement Control Association (AMCA) has established a standard testing method to ensure that published results are repeatable in the field. Called the *performance curves* by most manufacturers, these curves usually include the following:

• Static pressure or total pressure
• Brake horsepower
• Static efficiency or mechanical efficiency (also called total efficiency)

Figure C-29 shows an example of a performance curve.

A properly designed fan system would have a normal operating point where the static pressure or total pressure is at a peak.

With the damper closed, pressure measured on the discharge of a fan is just static pressure. Once the damper is opened, the static and velocity pressure can be measured. The *velocity pressure* refers to the pressure available for moving the gas. Velocity pressure cannot

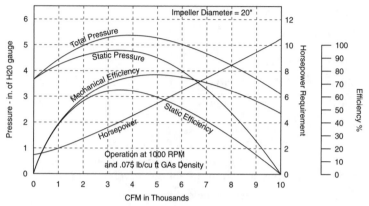

Figure C-29 Fan performance curve.

be measured with standard pressure gauges but instead requires the use of a Pitot tube (pronounced pea-toe). A Pitot tube was developed by French physicist Henri Pitot in 1771 to measure fluid velocity. A tube with a short angle bend is placed in a gas stream and a pressure is measured using a manometer or other accurate pressure-sensing device.

Total pressure is the sum of static pressure and velocity pressure. A Pitot tube and manometer arrangement is set up to measure the difference between the total pressure and the static pressure.

Almost all the pressure leaving the fan blades is velocity pressure. As the gas exits the fan housing, some of the velocity pressure is converted to static pressure. The conversion of velocity pressure to static pressure also occurs if the duct cross-sectional area increases.

System Curves

Gas transport systems consist of the fan, the inlet duct work, and the outlet duct work. A system can become more complicated with addition of dampers, cooling or heating coils, filters, diffusers, and noise attenuators.

The resistance to flow caused by all the components in system can be estimated using standard tables. One such table is the American Society of Heating, Refrigerating, and Air-Conditioning Engineers (ASHRAE) *Handbook of Fundamentals* (Atlanta: ASHRAE, 1997). Once the designer has determined the desired pressure and flow, the system curve can be generated. If the flow rate changes, the resulting resistance (pressure loss) to the flow will also change. The following relationship governs this change:

$$\frac{\text{Pressure}_2}{\text{Pressure}_1} = \left(\frac{\text{CFM}_2}{\text{CFM}_1}\right)^2$$

For example, if the system design point is determined to be 4800 CFM at a static pressure of 4.6 in of H_2O, other points on the design curve can be calculated as follows:

Point A - flow rate of 5300 CFM:

$$\frac{\text{Pressure}_2}{4.6} = \left(\frac{5300}{4800}\right)^2$$

Solving for Pressure$_2$

$$\text{Pressure}_2 = \left(\frac{5300}{4800}\right)^2 \times 4.6 = 5.6 \text{ in of } H_2O$$

5300 CFM @ 5.6 in of H_2O

Point B - flow rate of 2500 CFM:

$$\frac{\text{Pressure}_2}{4.6} = \left(\frac{2500}{4800}\right)^2$$

Solving for Pressure$_2$

$$\text{Pressure}_2 = \left(\frac{2500}{4800}\right)^2 \times 4.6 = 1.24 \text{ in of } H_2O$$

2500 CFM @ 1.24 in of H_2O

The system design point and the two calculated points can be plotted to determine the system curve, as shown in Figure C-30.

Once the system curve is developed, it can be plotted on the fan performance curve. The static pressure and brake horsepower curves are plotted in Figure C-31 along with the system curve (Point 1 in Figure C-31).

Note that the horsepower requirement is determined by moving straight down from the operating point to the BHP curve. The required horsepower is 5.5 HP. Most likely, a standard 7.5-HP motor will be used for this application.

Changes in the System

If, for some reason, the system characteristics are not as predicted, or a outlet damper is closed downstream, the system curve will change. Curve 2 shows a condition when the flow rate decreases to 3700 CFM. The fan static pressure will go up to 4.8 in of H_2O for

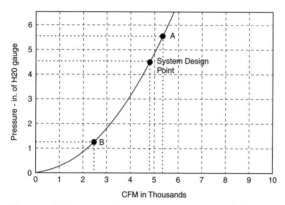

Figure C-30 System curve.

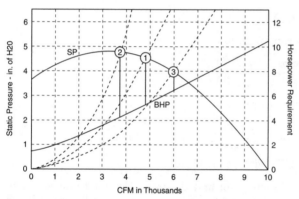

Figure C-31 Fan performance and system curve.

this condition and the horsepower requirement will drop to about 4.4 HP.

Curve 3 shows a condition when the flow rate increase to 6000 CFM. The fan static pressure will go down to 4.4 in of H_2O for this condition, and the horsepower requirement will increase to about 6.6 HP.

Changes in Fan Speed
Suppose it became necessary to get 10 percent more out of this same fan under design conditions. Then, for this example:

$$CFM2 = CFM1 \times 1.1 = 4800 \text{ CFM} \times 1.1 = 5280 \text{ CFM}$$

Increase in capacity could be achieved by speeding up the motor. This speed change will probably be accomplished with a change in sheaves on the belt drive. According to the fan laws, flow rate change is directly proportional to speed change:

$$\frac{CFM_2}{CFM_1} = \frac{RPM_2}{RPM_1}$$

The speed change requirement is calculated as follows:

$$\frac{5280}{4800} = \frac{CPM_2}{1000\,RPM}$$

Multiplying both sides by 1000 RPM,

$$CPM_2 = \frac{5280}{4800} \times 1000\,RPM = 1100\,RPM$$

NOTE

The fan curve is for 1000 RPM.

The new static pressure can be calculated using the fan laws as well:

$$\frac{SP_2}{SP_1} = \left(\frac{RPM_2}{RPM_1}\right)^2$$

For a $SP_1 = 4.7$ in. of H_2O (from the chart)

$$\frac{SP_2}{4.7\,\text{in. of}\,H_2O} = \left(\frac{1100}{1000}\right)^2$$

Multiplying both sides by 4.7 in. of H_2O,

$$SP_2 = \left(\frac{1100}{1000}\right)^2 \times 4.7\,\text{in. of}\,H_2O = 5.7\,\text{in. of}\,H_2O$$

Finally, the new horsepower requirement can be calculated:

$$\frac{BHP_2}{BHP_1} = \left(\frac{RPM_2}{RPM_1}\right)^3$$

$BHP_1 = 5.5\,HP$

$$\frac{BHP_2}{5.5HP_1} = \left(\frac{1100}{1000}\right)^3$$

Multiplying both sides by 5.5 HP,

$$\mathrm{BHP}_2 = \left(\frac{1100}{1000}\right)^3 \times 5.5\,\mathrm{HP} = 7.3\,\mathrm{HP}$$

This is good news. The original 7.5-HP motor does not have to be changed. These calculated changes are plotted in Figure C-32.

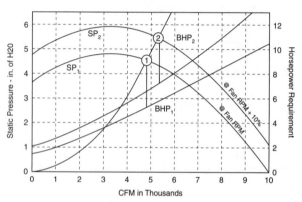

Figure C-32   System and performance curve 10 percent change in fan speed.

Note that the system curve has not changed. Nothing in the fan system has changed except the speed of the fan. However, this change in speed shifts the static pressure curve and the brake horsepower curves upward and to the right.

Changes in Density

Another change that may occur is a change in density of the gas being moved from the design density. Using the fan laws, a new static pressure can be calculated if the density increases to 0.1 lbs/ft³:

$$\frac{\mathrm{SP}_2}{\mathrm{SP}_1} = \frac{d_2}{d_1}$$

$\mathrm{SP}_1 = 4.6$ in of H_2O $d_2 = 0.1$ lbs/ft³

From the design curve $d_1 = 0.075$ lbs/ft³

$$\frac{\mathrm{SP}_2}{4.6} = \frac{0.1}{0.075}$$

Multiplying both sides by 4.6 in of H_2O,

$$SP_2 = \frac{0.1}{0.075} \times 4.6 = 6.1 \, \text{in of} \, H_2O$$

The chart in Figure C-33 shows the effect of this change.

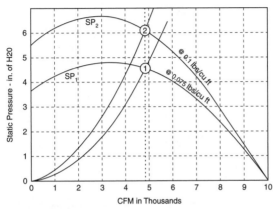

Figure C-33   Static pressure change with change in gas density.

Notice that both the fan performance and the system curve change. This is because the flow rate must remain the same under changing density conditions.

Glossary

activity — A task, job, or work order required in the completion of a project. Commonly used in Critical Path Method (CPM) scheduling, identified with a line and an arrow with a short description or abbreviation of the task. An activity can also be represented by a box in a Critical Path Method (CPM) network. See also *Critical Path Method*.

addendum — A document explaining a change or correction to a contract.

aggregate — Small stones or gravel used in construction of mortar or concrete. An aggregate of stone is also used on built-up roofs to dissipate heat.

allocation scheduling method — A technique that refocuses the workforce on preventive maintenance (PM) and predictive maintenance (PDM) work, while providing a facility-wide priority system to complete the other important jobs in the facility.

American National Standards Organization — An organization that publishes references defining the methods, classification, and testing of materials and standard languages used in science and industry. See also *American National Standards Institute*.

American National Standards Institute (ANSI) — An organization that develops trade and communications standards.

Arrow Diagram Method (ADM) — The traditional or first method for representing a logic network by identifying an activity as an arrow with circles (or events), noting its beginning and ending.

as late as possible (ALAP) — In Critical Path Method (CPM) and Project Evaluation and Review Technique (PERT), designates a task that should start as late as possible, using all the free-float time available. See also *Critical Path Method* and *Project Evaluation and Review Technique*.

as soon as possible (ASAP) — In Critical Path Method (CPM) and Project Evaluation and Review Technique (PERT), indicates a task that should start as soon as possible, given the start date of the project and its predecessor tasks. See also *Critical Path Method* and *Project Evaluation and Review Technique*.

backlog age — A measurement of the number of work orders in the backlog that can be completed within their priority period and the number that has missed the deadline.

backlog hours — The number of direct labor hours identified by work orders currently on hand.

battery limits — Area around equipment, which is usually confined.

benchmark (in slotting) — A short description of a job and an estimate of the labor hours required to complete it. The description and estimate are usually laid out on a spreadsheet by job type or craft, and in order of labor hours required.

benchmark job — A short description of a job and an estimate of the labor hours required to complete it. The description and estimate are usually laid out on a spreadsheet by job type or craft, and in order of labor hours required.

computerized maintenance management system (CMMS) — A computer system that tracks the relationship between work orders and associated records.

CPM — See *Critical Path Method*.

crash time — The shortest duration of the project.

critical path — The series of jobs in a critical path network that make up the longest path to the completion of a project. Alternatively can be the sequential list of jobs from the beginning to the end of a critical path network, having no float. See also *Critical Path Method* and *float*.

Critical Path Method (CPM) — A system of project scheduling used to identify the sequence of activities and milestones required to complete a project.

critical path network — Used in Critical Path Method (CPM), the complete diagram of all the activities and milestones required to complete a project. See also *Critical Path Method*.

earliest completion time — The calculation of the earliest possible finish time of each activity. Calculated by starting at the beginning event and adding up subsequent activity times.

earned value — A performance measure calculated by multiplying a task's planned cost by the percentage of work completed. Required in all projects for the U.S. government.

elapsed time — The total time required to complete a job or task. Elapsed time is equal to the labor hours if only one person is assigned to the job. If more than one person is assigned to the job, the elapsed time will be less than the labor hours.

event — Used in Critical Path Method (CPM), arrow diagram to describe the starting point of one activity and the beginning point of another activity. An event will be indicated in a critical path network diagram by a number enclosed in a circle. See also *Critical Path Method* and *critical path network*.

expediting — Processing of *rush* orders that are needed in less than normal lead time, the follow-up activity for orders that are overdue, or checking on important orders to determine status and progress.

float — The amount of time a noncritical task can be delayed before it influences another task's schedule. Also called *slack*.

Gantt chart — A graphical representation of a project schedule that shows each task as a bar whose length is proportional to its duration. Many programs include a spreadsheet section to the left of the Gantt chart to display a selection of project data.

hammock task — In Critical Path Method (CPM) and Project Evaluation and Review Technique (PERT), a task whose duration is calculated based on the time span between its predecessor and successor activities. Figuratively suspended between other tasks. See also *Critical Path Method* and *Project Evaluation and Review Technique*.

histogram — In Critical Path Method (CPM) and Project Evaluation and Review Technique (PERT), a bar chart that shows resource workloads by time period. See also *Critical Path Method* and *Project Evaluation and Review Technique*.

labor hours — The total hours worked by all individuals on a job or task. Formerly referred to as man-hours.

lag — In Critical Path Method (CPM) and Project Evaluation and Review Technique (PERT), the amount of time between the finish of a predecessor task and start of a successor task. See also *Critical Path Method* and *Project Evaluation and Review Technique*.

latest completion times — The latest possible times that each activity can be finished without increasing the length of the project. Determined by starting with the total project elapsed time at the end of the critical path network and subtracting activity times until the first activity is reached. See also *critical path network*.

lead — In Critical Path Method (CPM) and Project Evaluation and Review Technique (PERT), the amount of time that a successor task is permitted to start before its predecessor is finished. See also *Critical Path Method* and *Project Evaluation and Review Technique*.

load leveling — The distribution of resources so that constraints on the resources are not violated.

man-hours — See *labor hours*.

Methods Time Measurement (MTM) — The development of job steps based on the basic movements of a human's hands, feet, eyes, and body. Times associated with these movements are measured down to the nearest 0.00001 hours.

milestone — A project event that represents a checkpoint, a major accomplishment, or a measurable goal. Also, a significant point in the development of a project or job. Commonly used in Critical Path Method (CPM). See also *Critical Path Method*.

negative float (negative slack) — In Critical Path Method (CPM) and Project Evaluation and Review Technique (PERT), unscheduled delay before an actual task starts. This is time that must be recovered if the project is not to be delayed. See also *Critical Path Method* and *Project Evaluation and Review Technique*.

Organizational Breakdown Structure (OBS) codes — A set of codes used in project programs to identify tasks by resource groups in a hierarchy. Often used to reflect departmental structure in a company or code of accounts. This structure can be used for filtering tasks.

PERT — See *Project Evaluation and Review Technique*.

PERT chart — A graphical depiction of task dependencies, resembling a flowchart. Dependencies are indicated by connecting lines or arrows that show the work flow. Also called a *network diagram*.

planning — The allocation of needed resources and the sequence in which they are needed to allow an essential activity to be performed in the shortest time or at the least cost.

planning thought process — A method that uses the abilities of a planner to the fullest to establish an estimate of time required for the job, define understandable steps to complete the job, and identify the materials, parts, tools, and equipment required for the job.

Precedent Diagram Method (PDM) — A method for dispensing with event identification and instead using a box to represent the activity. Lines (drawn from left to right) represent the interdependencies of different activities.

precedent logic — The method of putting activities in sequential order.

predecessor — In Critical Path Method (CPM) and Project Evaluation and Review Technique (PERT), the task that must be started or completed first in a dependency relationship between two tasks. See also *Critical Path Method* and *Project Evaluation and Review Technique*.

predictive maintenance (PDM) — A program that compares test measurements to established engineering limits in order to determine the need for corrective work. The limits are set to ensure that sufficient time is available for repair and to prevent an emergency shutdown of the equipment.

preventive maintenance (PM) — The maintenance work performed during maintenance custody on equipment through manufacturer-recommended rebuilds or rebuilds required because of predictable wear. Also, the maintenance work performed during production custody on equipment (including adjustments, lubrication, tests, inspections, and calibrations).

prioritization — A process that begins as requests for maintenance work are received and breaks down the requests into three basic stages: classification, requested completion date, and schedule priorities.

Project Evaluation and Review Technique (PERT) — A method of reviewing projects similar to the Critical Path Method, except that it also incorporates the uncertainty associated with completing each task in the project. A PERT diagram can now also (erroneously) refer to the network diagram developed through Critical Path Method (CPM) in many computerized project programs. This new definition has had growing acceptance because of the proliferation of project programs, even though it has little to do with the original intent of the method.

resource — Any person, group of people, piece of equipment, or material used in accomplishing a task or job.

resource-driven — Task durations determined by a project program based on the number and allocation of resources, rather than the time available. Both tasks and entire projects can be resource-driven.

resource-leveling — The process of resolving resource conflicts.

shutdowns — Down periods used for inspections in organizations. Also referred to as a *shut-in*, a *downturn*, a *turnaround*, or an *outage*.

slack — See *float*.

slotting — Positioning of an MTM-based estimated benchmark job on a spreadsheet. Slotting is also referred to as choosing a job from the spreadsheet that is the same or similar to a job to be estimated. See also *Methods Time Measurement* and *benchmark*.

square foot cost estimating — An estimating method used to prepare budget or preliminary estimates for construction projects.

subproject — A group of activities treated as a single task in a master project schedule. Subprojects are a way of working with multiple projects by keeping all the data in one file rather than independent files.

successor — A task that must await the start or completion of another in a dependency relationship between two tasks.

take-off — Bill of material or document that records the dimensions and quantities to be included in the estimate.

unit cost — The value assigned to a single unit of work. The total cost is determined when the unit cost is multiplied by a quantity.

vibration analysis — An analysis of vibration in rotating machinery to discover problems such as misalignment, imbalance, bearing damage, and looseness.

Work Breakdown Structure (WBS) codes — A set of codes in project programs used to identify tasks in a hierarchy. Many programs associate these codes with an outline structure. The WBS codes can be used to filter the schedule.

work order — A formal means of requesting maintenance services, making it the focus of most maintenance records. Provides a financial structure to the work that the Maintenance department performs.

work package — A folder (or group of folders) that includes all the necessary paper work and reference material required to complete a job. The work order, written procedure, specification sheets, tool lists, parts lists, sketches, prints, permits, and equipment manuals should be included.

Index